Lecture Notes in Mathematics

Edited by A. Dold and B. Eckmann
Series: Australian National University, Canberra
Advisers: L. G. Kovács, B. H. Neumann and M. F. Newman

573

Group Theory

Proceedings of a Miniconference
Held at the Australian National University,
Canberra, November 4–6, 1975

Edited by R. A. Bryce, J. Cossey and M. F. Newman

Springer-Verlag
Berlin · Heidelberg · New York 1977

Editors
R. A. Bryce
J. Cossey
Department of Pure Mathematics
Faculty of Arts
Australian National University
Canberra ACT 2600/Australia

M. F. Newman
Department of Mathematics
Institute of Advanced Studies
Australian National University
Canberra ACT 2600/Australia

AMS Subject Classifications (1970): 05D25, 20C05, 20C15, 20D10, 20D15, 20D45, 20E15, 20E25, 20E99, 20H15, 20J99, 20K99

ISBN 3-540-08131-3 Springer-Verlag Berlin · Heidelberg · New York
ISBN 0-387-08131-3 Springer-Verlag New York · Heidelberg · Berlin

Printed in Germany
Printing and binding: Beltz Offsetdruck, Hemsbach/Bergstr.
2141/3140 543210

INTRODUCTION

The impetus for this Miniconference came from the conjunction of visits to Canberra of Tom Berger and John Wilson. This seemed a good opportunity to correct the regretably rare occurrence of meetings on specialized topics by having a meeting of group theorists from all over Australia. At very short notice it was possible to stage a conference with over 40 participants. Our thanks go to the Australian National University and to the other Universities involved for their financial support.

Initially we advertised the conference topic as 'soluble groups', but the number of people eager to attend with contributions from general group theory was so great that we generalized the title.

The conference consisted of five half-day sessions beginning with a one hour invited lecture followed by contributions of length varying from 20 minutes to an hour. One only of the invited lectures is not included. L.G. Kovács invited talk - "Permutation groups of prime degree" - was an account of a proof by J.A. Green of a result of Peter Neumann. This result has been extended by M. Klemm in a paper entitled "Über einen Satz von P.M. Neumann" to appear in *Communications in Algebra.*

LIST OF PARTICIPANTS

Mr Q. Abbasi, Australian National University
Mrs J.A. Ascione, Australian National University
Miss B. Bailey, Australian National University and University of Minnesota
Professor T.R. Berger, Australian National University and University of Minnesota
Professor R.G. Burns, University of Queensland and York University, Canada
Dr R.A. Bryce, Australian National University
Dr J.J. Cannon, University of Sydney
Mr C. Charnes, Monash University
Mr G. Cole, Rusden State College of Victoria
Dr S.B. Conlon, University of Sydney
Dr P.J. Cossey, Australian National University
Mr B. Dolman, University of Adelaide
Dr T.M. Gagen, University of Sydney
Dr J.R.J. Groves, University of Melbourne
Dr G. Havas, Australian National University
Dr M. Herzog, Australian National University
Mr R.B. Howlett, University of Adelaide
Dr G. Ivanov, University of Sydney
Dr G. Karpilovsky, University of New South Wales
Mr A.C. Kim, Australian National University
Dr L.G. Kovács, Australian National University
Dr H. Lausch, Monash University
Dr G. Lehrer, University of Sydney
Dr D. McDougall, University of Tasmania
Dr M.F. Newman, Australian National University
Mr G. Pain, Monash University
Dr D. Parrott, University of Adelaide
Dr C.E. Praeger, Australian National University
Dr A.J. Rahilly, Gippsland College of Advanced Education
Dr K.M. Rangaswamy, Australian National University
Dr H. Sarges, University of Sydney
Mr D.B. Shield, Australian National University
Dr H.L. Silcock, University of Adelaide,
Dr J.B. Southcott, Flinders University
Dr D.E. Taylor, University of Sydney
Dr J.W. Wamsley, Flinders University
Dr J.N. Ward, University of Sydney
Dr J.S. Wilson, Australian National University and University of Cambridge, England
Mr W.H. Wilson, University of Sydney
Dr David Wright, Australian National University
Dr Donald Wright, Monash University

CONTENTS

* denotes an invited lecture.

PROC. MINICONF. THEORY OF GROUPS
CANBERRA 1975, 1-5.

20D45
(20D10)

FIXED POINT FREE AUTOMORPHISM GROUPS

T.R. Berger

Frobenius conjectured that if A is a cyclic operator group of prime order p which acts as a fixed-point-free operator group of a group G (that is, $C_G(A) = 1$) then G is nilpotent. In unpublished work, Witt (1935) showed that if G is solvable then G is nilpotent. The proof that G is solvable was given by Thompson in 1959 [9]. Since then, the conjecture has expanded in its scope.

CONJECTURE. *If* A *is a finite solvable fixed-point-free operator group of a finite group* G *where* $(|A|, |G|) = 1$ *then*

(1) G *is solvable, and*

(2) *the Fitting (nilpotent) length of* G *is bounded above by the composition length of* A .

A history of this conjecture is contained in [5] with references. In this discussion, in referring to the conjecture, we shall only mean part (2). Results of the mid-1960's are numerous and often overlap due to the various extra hypotheses required in the statement of theorems. From the many results, however, one may be singled out for the power of its methods: that is, the paper [8] of Shult. This paper may serve as a good beginning for any study of part (2) of the conjecture.

Let us begin a proof of the conjecture now. Proceeding via a counter-example which minimizes $|G|$, we may show that the Fitting subgroup M of G is a unique minimal normal A-invariant subgroup of G . Consequently, $G = G_0M$ is a split extension of M by an A-invariant subgroup G_0 . If M is a p-group then we may view M as a faithful irreducible $GF(p)$-module for the semidirect product AG_0 . With a little more arguing we may even assume that M is an absolutely irreducible

$K[AG_0]$-module for an extension K of $GF(p)$.

We have reduced the counter example to the following configuration:

(*) AG_0 *acts faithfully and absolutely irreducibly on a* $K[AG_0]$*-module* M.

In addition, both $C_M(A) = (0)$ and $C_{G_0}(A) = 1$. This latter condition $(C_{G_0}(A) = 1)$ is not an easy one to deal with directly, so it is this one which we relax. The idea is to devise a new hypothesis H which, together with (*), is implied by the conditions of the conjecture. These new conditions should imply that $C_M(A) \neq (0)$. Clearly such a theorem would give a proof to part (2) of the conjecture.

Generally, the hypothesis H will have to be quite involved, as will the conclusion of an appropriate theorem (for example, see [1, Theorem B]). The intricacies of H, along with the conclusion, are necessary to skirt fine details in proof. In order to make things clear, let us ignore all detail and state things in clear (but incorrect) form. We know little about M except that, for a counter-example, $C_M(A) = (0)$, that is $M|_A$ does not contain a trivial A-submodule. Suppose that N is a $K[AG_0]$-module which does not contain a copy of a nontrivial A-module V. If $\hat{V}$ is the dual of V viewed as a $K[AG_0/G_0]$-module then $N \otimes_K \hat{V}$ does not contain the trivial A-submodule. In other words, any composition factor of $N \otimes_K \hat{V}$ might serve for M. So long as $N|_A$ is missing modules, an M such that $C_M(A) = (0)$ can exist. Thus our theorem should imply that $M|_A$ contains all irreducible $K[A]$-modules. It turns out that this condition is little different from asking that $M|_A$ have a regular A-direct summand. We are now in a position to state a hypothetical theorem.

APPROXIMATE THEOREM. *Let* A *be a finite solvable operator group of a finite solvable group* G_0 *where* $(|A|, |G_0|) = 1$. *Let* K *be an algebraically closed field of characteristic* p *where* $p \nmid |A|$, *and let* M *be a faithful irreducible* $K[AG_0]$*-module. If the triple* (A, G_0, M) *satisfies some suitable additional hypothesis* H *then* $M|_A$ *has a regular direct summand.*

In order to prove the theorem one must analyze $M|_A$, a method for doing this being given in [4, §5]. For some subgroup A_1G_1, where $A_1 \leq A$ and $G_1 \leq G_0$, of AG_0, and some primitive $K[A_1G_1]$-module M_1, $M \simeq M_1|^{AG}$. We assume that the theorem is false and choose a counterexample minimizing $|A| + |G_0| + \dim M$. The hypothesis H must be chosen so that it alone implies that $A_0 = A$. If we set

$M_1^* = M_1|^{A_1G_0}$ then $M \simeq M_1^*|^{AG_0}$ where G_0 has little to do with this induction step. The possibility of verifying the hypothesis for (A_1, G_0, M_1^*) and then using the inductive hypothesis to prove the theorem, rests upon the strength of H.

The next object is to prove that $G_1 = G_0$ so that M is a primitive module. Assuming the contrary, choose $G_2 \geq G_1$ an A-invariant maximal subgroup of G_0, and an A-chief factor $\overline{H} = H/K$ of G_0 such that $G_0 = G_2H$ and $K = G_2 \cap H$. Setting $M_2 = M_1|^{AG_2}$, we may use Mackey's induction theorem to show that

$$M|_A \simeq M_2|^{AG}|_A \simeq \sum^{\oplus} \left(x^{-1} \otimes M_2\right)|_{A\cap(AG_2)^x}|^A$$

where the sum is over A, AG_2-double cosets in AG_0. Since $G_0 = G_2H$ and $K = G_2 \cap H$ the elements x may be chosen so that xK runs over a complete set of A-orbit representatives for the conjugation action of A upon $\overline{H}$. Further, in the coset xK, since $(|A|, |G_0|) = 1$, we may choose x so that $A \cap (AG_2)^x = C_A(x)$. In other words,

$$x^{-1} \otimes M_2|_{C_A(x)} \simeq M_2|_{C_A(x)} .$$

Continuing, we have

$$M|_A \simeq \sum^{\oplus} M_2|_{C_A(x)}|^A .$$

Using the inductive hypothesis it is possible to verify H for $(C_A(\overline{H}), G_2, M_2)$. In particular, $M_2|_{C_A(\overline{H})}|^A$ must have a regular A-direct summand. We want $C_A(\overline{H}) = C_A(x)$ for some one of the x's. But $C_A(x)$ is the point stabilizer in A for the conjugation action of A upon $\overline{H}$. The "good" x will only exist if $A/C_A(\overline{H})$ has a regular orbit upon $\overline{H}$.

QUESTION 1. *If AG_2 acts irreducibly upon a module $\overline{H}$, then does $A/C_A(\overline{H})$ have a regular orbit in its action upon $\overline{H}$?*

If the answer is "yes" then the assumption that $G_1 < G_0$ is false completing the proof that $A_1G_1 = AG$ so that M is a primitive module.

The answer to Question 1 is not always "yes". For example, if $A = D_8 \simeq Z_2 \wr Z_2$ then A has a faithful absolutely irreducible module $\overline{H}$ over GF(3) of dimension

2 (order 9) such that A has no regular orbits upon the elements of $\overline{H}$. On the other hand, if H includes the condition that A be nilpotent and $Z_p \wr Z_p$-free for all primes p then the question always has an affirmative answer.

Since M is primitive, we know a good deal about the structure of the Fitting subgroup $F(G_0)$. Arguing further on tensor decompositions of M into a product of projective $K[AG_0]$-modules, we may actually assume that

(**) *$F(G_0) = R$ is an extra special r-group, $Z(R) = Z(G_0)$, $R/Z(R)$ is a chief factor of AG_0 , and $M|_R$ is faithful and irreducible.*

QUESTION 2. *If (**) holds then does $M|_A$ have a regular A-direct summand?*

If the answer is "yes", the proof of our theorem is complete. Again, if H includes the condition that A be nilpotent and $Z_p \wr Z_p$-free then Question 2 has an affirmative answer unless $G_0 = R$. This final possibility $(G_0 = R)$ may be avoided by adjusting H and the conclusion of our theorem.

As an example, consider $SL(2, 3) = AG_0$ where A is cyclic of order 3 and G_0 is quaternion of order 8 . Now AG_0 has a faithful absolutely irreducible $K[AG_0]$-module M of dimension 2 over K of characteristic $p \neq 3$ or 2 such that $M|_A$ is the sum two distinct nontrivial irreducible $K[A]$-modules. Certainly, $M|_A$ does not have a regular A-direct summand.

It is clear that a proof of our theorem centers on consideration of Questions 1 and 2. These questions first received serious attention in the landmark paper [7] of Hall and Higman. Thus one wants to consider a cyclic p-subgroup A normalizing a subgroup G_0 inside the linear group $GL(M)$ where M is a vector space over a field K of characteristic p . If $A = \langle x \rangle$ has order p^n then under what conditions is the minimal polynomial of x on V equal to $X^{p^n} - 1$? In an equivalent way, when does $V|_A$ have a regular A-direct summand? One reduces to the case where G_0 is a special group on which A acts faithfully. From here the argument exactly parallels that of our theorem. The importance of Question 1 is somewhat hidden since A will act semiregularly upon the elements of $\overline{H}$ since A is cyclic. But the Question 2, where $G_0 = F(G_0) = R$, is the classic one whose solution is often just called "Theorem B".

If we allow K to be of any characteristic, we see that Questions 1 and 2 are important in their own right. Thus consideration of these two questions is the core of the Hall-Higman series of papers.

Questions 1 and 2 are really quite similar: one asks for regular orbits; the other for regular direct summands. Consequently, the methods evolved to handle these two questions are quite similar. These methods are discussed in [4, §5] and are much more generally applicable than just to the conjecture. In summing up, the original conjecture led to a consideration of Questions 1 and 2, whose breadth is greater than just the original conjecture. In turn, consideration of these two questions led to methods for analyzing completely reducible representations of solvable groups. These methods go far beyond even the two questions. To give some idea as to the breadth of applicability of these methods, the following papers are cited [2, 3, 5, 6].

REFERENCES

[1] T.R. Berger, "Automorphisms of solvable groups", *J. Algebra* **27** (1973), 311-340. MR**50**#476.

[2] Thomas R. Berger, "Nilpotent fixed point free automorphism groups of solvable groups", *Math. Z.* **131** (1973), 305-312. MR**49**#2940.

[3] T.R. Berger, "Characters and derived length in groups of odd order", *J. Algebra* (to appear).

[4] T.R. Berger, "Hall-Higman type theorems V", *Pacific J. Math.* (to appear).

[5] T.R. Berger, "Irreducible modules of solvable groups are algebraic", *Proc. Park City Group Theory Conf.*, 1975 (to appear).

[6] T.R. Berger and F. Gross, "2-length and derived length of a Sylow 2-subgroup", *Proc. London Math. Soc.* (to appear).

[7] P. Hall and Graham Higman, "On the p-length of p-soluble groups and reduction theorems for Burnside's problem", *Proc. London Math. Soc.* (3) **6** (1956), 1-42. MR**17**,344.

[8] Ernest E. Shult, "On groups admitting fixed point free abelian operator groups", *Illinois J. Math.* **9** (1965), 701-720. MR**32**#1269.

[9] John Thompson, "Finite groups with fixed-point-free automorphisms of prime order", *Proc. Nat. Acad. Sci. USA* **45** (1959), 578-581. MR**21**#3484.

Department of Pure Mathematics,
Faculty of Arts,
Australian National University,
Canberra, 2600, Australia;

Permanent address: Department of Mathematics,
University of Minnesota,
Minneapolis,
Minnesota, USA.

PROC. MINICONF. THEORY OF GROUPS
CANBERRA 1975, 6-16. 20D10

STRONG CONTAINMENT OF FITTING CLASSES

R.A. Bryce and John Cossey

1. Introduction

A Fitting class $\underline{\underline{X}}$ is said to be strongly contained in a Fitting class $\underline{\underline{Y}}$ if in every group G an $\underline{\underline{X}}$-injector is always contained in a $\underline{\underline{Y}}$-injector (see Hartley [4] for definitions, and note that groups are always finite and soluble). We denote strong containment of $\underline{\underline{X}}$ in $\underline{\underline{Y}}$ by $\underline{\underline{X}} << \underline{\underline{Y}}$.

Although strong containment clearly implies containment, it is easy to show the converse is not true. For example, an $\underline{\underline{S}}_2$-injector of S_3 , the symmetric group on three letters, is a Sylow 2-subgroup, while the Sylow 3-subgroup is an $\underline{\underline{N}}$-injector, giving $\underline{\underline{S}}_2$ not strongly contained in $\underline{\underline{N}}$; clearly $\underline{\underline{S}}_2 \subseteq \underline{\underline{N}}$.

There are two cases we know of where strong containment arises from a general dispensation. Since every π-subgroup of a (finite soluble) group is contained in a Hall π-subgroup, which is a S_π-injector, $\underline{\underline{X}} \subseteq \underline{\underline{S}}_\pi$ implies $\underline{\underline{X}} << \underline{\underline{S}}_\pi$. The other example is that of strong normality (we refer the reader to §§2, 3 of [2] for definitions and results). Lockett [6] has shown that strong normality implies strong containment; in §2 we give an alternative proof of this.

Our impetus for studying strong containment came from a paper of Makan [7]: we are grateful to him for providing us with a preprint of the paper. There he characterizes normal Fitting classes in terms of strong containment, showing, *inter alia*, that $\underline{\underline{X}}$ is normal if and only if $\underline{\underline{X}} \cap \underline{\underline{N}}^i << \underline{\underline{N}}^i$ for all positive integers i . Our basic observation is that the machinery Makan sets up can be used to prove with very little extra effort that the Fitting classes strongly contained in each $\underline{\underline{N}}^i$ are precisely the strongly normal ones (this is carried out in §3).

Our attempts to go beyond this have not met with much success. We have been able to establish some results of a general nature which are useful in dealing with particular examples: these are collected in §4. In §5, we record our attempts to deal with small nilpotent length: even here, where individual cases are often easy to dispose of, we have not been able to make much progress.

2. An alternative proof of a result of Lockett

(2.1) THEOREM (Lockett [6]). *Let* $\underline{\underline{X}}$, $\underline{\underline{Y}}$ *be Fitting classes with* $\underline{\underline{X}}$ *strongly normal in* $\underline{\underline{Y}}$ *(that is,* $\underline{\underline{Y}}_* \subseteq \underline{\underline{X}}$ *).* $\underline{\underline{X}}$ *is strongly contained in* $\underline{\underline{Y}}$.

It will suffice to prove that if G is a group and if V is a $\underline{\underline{Y}}$-injector of G then $V_{\underline{\underline{X}}}$ (the $\underline{\underline{X}}$-radical of V) is an $\underline{\underline{X}}$-injector of G .

This result, it turns out, is very close to the existence of injectors at all, and uses a restatement of the crucial lemma in [3] where the existence of injectors was demonstrated.

(2.2) LEMMA. *Let* G *be a finite soluble group and* $\underline{\underline{X}}$ *a Fitting class. Suppose that* W *is a normal subgroup of* G *and that* V_1, V_2 *are maximal* $\underline{\underline{X}}$*-subgroups of* G *containing* W *such that*

$$V_i/W \leq \zeta\left(N_{G/W}(V_i/W)\right) , \quad i = 1, 2 .$$

Then V_1, V_2 *are conjugate.* (Here ζ denotes the hypercentre.)

The proof is easily obtained from that in [3]. The other fact we need is (2.1) (c) in [6].

(2.3) LEMMA. *If* $\underline{\underline{X}}$ *is strongly normal in* $\underline{\underline{Y}}$ *and* $G \in \underline{\underline{Y}}$ *then*

$$[G, \text{Aut } G] \leq G_{\underline{\underline{X}}} .$$

Proof of (2.1). We prove by induction on $|G|$ that if V is a $\underline{\underline{Y}}$-injector of the group G then $V_{\underline{\underline{X}}}$ is an $\underline{\underline{X}}$-injector of G . Suppose $|G| > 1$ and that the result is proved for groups of order less than $|G|$. Let N be a normal subgroup of G with $1 \neq G/N$ being nilpotent. Let U be an $\underline{\underline{X}}$-injector of G and choose V to be a maximal $\underline{\underline{Y}}$-subgroup of G containing U .

Now U is an $\underline{\underline{X}}$-injector of V so $V_{\underline{\underline{X}}} \leq U$. But $V/V_{\underline{\underline{X}}}$ is abelian by (2.3) so $U \trianglelefteq V$ whence $U = V_{\underline{\underline{X}}}$. In particular U is characteristic in V and it follows that $U \cap N$ is normalized by $N_G(V)$. Next,

(2.4) $\qquad V$ is hypercentral in $N_G(V)$ modulo $U \cap N$.

For, using (2.3), $V \cap N/U \cap N$ is centralized by $N_G(V)$ and

$$[V, \underbrace{N_G(V), \ldots, N_G(V)}_{r}] \leq V \cap N$$

for large enough r , since G/N is nilpotent.

Next observe that $U \cap N$ is an $\underline{\underline{X}}$-injector of N . Let T be a $\underline{\underline{Y}}$-injector of G so that $T \cap N$ is a $\underline{\underline{Y}}$-injector of N . It follows from the inductive hypothesis that $U \cap N$ is conjugate to $(T \cap N)_{\underline{\underline{X}}}$. Without loss of generality suppose T chosen so that $U \cap N = (T \cap N)_{\underline{\underline{X}}}$. Then $U \cap N$ is normalized by T . As above

(2.5) T is hypercentral in $N_G(T)$ modulo $U \cap N$.

Put $H = \langle T, V \rangle$. Then $U \cap N \trianglelefteq H$ and T, V satisfy the conditions of (2.2): both are maximal $\underline{\underline{Y}}$-subgroups of H and by (2.4), (2.5) each is hypercentral in its normalizer in H modulo $U \cap N$. Hence T, V are conjugate, so V is indeed a $\underline{\underline{Y}}$-injector of G . Since $U = V_{\underline{\underline{X}}}$ the induction is complete.

3. Strong containment in $\underline{\underline{N}}^k \cap \underline{\underline{S}}_\pi$

We aim to prove here the result alluded to in §1.

(3.1) THEOREM. *Let* π *be a set of primes containing either one or at least three elements, and let* k *be a positive integer. If* $\underline{\underline{X}}$ *is a Fitting class strongly contained in* $\underline{\underline{N}}^k \cap \underline{\underline{S}}_\pi$ *then* $\underline{\underline{X}}$ *is strongly normal in* $\underline{\underline{N}}^k \cap \underline{\underline{S}}_\pi$.

The crux of the proof is this result of Makan's [7].

(3.2) LEMMA. *Let* π *be a set of primes of cardinality* 1 *or at least* 3 *, and let* k *be a positive integer. Suppose that* $\underline{\underline{X}}$ *is a non-trivial Fitting class strongly contained in* $\underline{\underline{N}}^k \cap \underline{\underline{S}}_\pi$. *If* $G \in \underline{\underline{X}} \cap \underline{\underline{N}}^{k-1} \cap \underline{\underline{S}}_\pi$ *and* $p \in \pi$ *then there is a positive integer* m *such that for all* n ,

$$G^{(mn)} \text{ wr } C_p \in \underline{\underline{X}} .$$

A convenient result is

(3.3) LEMMA. *Let* $\underline{\underline{X}}, \underline{\underline{Y}}$ *be Fitting classes with* $\underline{\underline{X}} \subseteq \underline{\underline{Y}}^*$ *and let* q *be a prime. We suppose that* $G \in \underline{\underline{Y}}^*$ *whenever* G *is a group whose nilpotent residual* $G^{\underline{\underline{N}}}$ *is an* $\underline{\underline{X}}$*-group complemented by a cyclic group of order* q *. Then*

$$\underline{\underline{X}}\underline{\underline{S}}_q \subseteq \underline{\underline{Y}}^* .$$

Proof. First we show that every split extension of an $\underline{\underline{X}}$-group by a cyclic group of order q is in $\underline{\underline{Y}}^*$. For, suppose $H = MD$ $(M \in \underline{\underline{X}}, |D| = q)$. Consider the descending chain of subgroups H_i of H defined by

$$H_0 = H ,$$

$$H_{i+1} = D^{H_i} , \quad i \geq 0 .$$

For some n , $H_n = H_{n+1} = \dots = K$ say. Now $K = (K \cap M)D$ and $K^{\underline{\underline{N}}} = K \cap M$. The latter statement comes as follows:

$$K \cap M = D^K \cap M = D[K, D] \cap M = [K, D]$$
$$= [(K \cap M)D, D] = [K \cap M, D]$$

whence for all t ,

$$K \cap M = [K \cap M, tD]$$

and so

$$K \cap M \leq K^{\underline{\underline{N}}} .$$

But plainly $K^{\underline{\underline{N}}} \leq K \cap M$ so $K^{\underline{\underline{N}}} = K \cap M$.

By hypothesis now $K \in \underline{\underline{Y}}^*$ and K is subnormal in H . It follows that $H \in N_0\{K, M\} \subseteq \underline{\underline{Y}}^*$ as required.

Suppose now that the conclusion of the lemma is false. Let $G \in \underline{\underline{XS}}_q \backslash \underline{\underline{Y}}^*$ have least order. Then G has a unique maximal normal subgroup and so $G/G^{\underline{\underline{N}}}$ is cyclic of q-power order. It follows that $G = G^{\underline{\underline{N}}}C$ where $C \cong C_{q^\beta}$ say. Put $A = G^{\underline{\underline{N}}}$. Form the group $\hat{G}$, the split extension of A by C using the action of C on A induced by conjugation in G . Let $\theta : \hat{G} \twoheadrightarrow G$ be the natural homomorphism. Now

$$A \cap \ker \theta = 1 .$$

It follows from [1, Lemma 1.1] that $\hat{G} \notin \underline{\underline{Y}}^*$. That is, if $\underline{\underline{XS}}_q \not\subseteq \underline{\underline{Y}}^*$ there is a split extension of an $\underline{\underline{X}}$-group by a q-power cyclic group not in $\underline{\underline{Y}}^*$. Choose such a group $L = BC$ of least order: $B \trianglelefteq L$, $B \cap C = 1$, $B \in \underline{\underline{X}}$, $C \cong C_{q^\alpha}$. Note that $\alpha > 1$ on account of the result above.

Now let C_0 be the subgroup of order $q^{\alpha-1}$ in C . Consider the group

$$W = BC_0 \text{ wr } D$$

where $D \cong C_q$. Since W is generated by the subnormal $\underline{\underline{Y}}^*$-subgroups $B \text{ wr } D$ and the base group, $W \in \underline{\underline{Y}}^*$.

Put $C = \langle c \rangle$, $D = \langle d \rangle$. Let $f \in \left(BC_0\right)^D$ be defined by

$$f(1) = c^q , \quad f\left(d^i\right) = 1 , \quad 1 \leq i \leq q-1 .$$

Put $\beta = fd$. The subgroup $B^D\langle\beta\rangle \in S_n\{W\} \subseteq \underline{\underline{Y}}^*$. Note that

$$\beta^q = \prod_{i=1}^{q} f^{d^i} . \tag{3.4}$$

The group C acts on B^D in a natural way, namely component wise: for all $g \in B^D$

$$g^c\left(d^i\right) = g\left(d^i\right)^c , \quad 0 \le i \le q-1 .$$

The actions of β and C on B^D commute; and since by (3.4) β^q and c^q agree in their action on B^D we have $\left(c^{-1}\beta\right)^q$ acts trivially on B^D . Consequently if we form the split extension

$$W_1 = B^D\langle c, \beta\rangle$$

the subgroup $U_1 = \langle \left(c^{-1}\beta\right)^q \rangle$ centralizes B^D and is therefore normal. Since $W_1/U_1 \in \underline{\underline{Y}}^*$ it follows from [1, Lemma 1] that $B^D\langle c^{-1}\beta\rangle \in \underline{\underline{Y}}^*$. Then

$$W_1 \in N_0\{B^D\langle\beta\rangle, B^D\langle c^{-1}\beta\rangle\} \subseteq \underline{\underline{Y}}^* ,$$

and finally $B^DC \in s_n W_1 \in \underline{\underline{Y}}^*$. This means $BC = L \in \underline{\underline{Y}}^*$, a contradiction to our assumption that $\underline{\underline{XS}}_q \not\subseteq \underline{\underline{Y}}^*$.

Proof of (3.1). We must show that under the hypotheses, $\underline{\underline{Y}}^* = \underline{\underline{N}}^k \cap \underline{\underline{S}}_\pi$. Suppose to the contrary that $\underline{\underline{Y}}^* \ne \underline{\underline{N}}^k \cap \underline{\underline{S}}_\pi$. This means (since $\underline{\underline{N}}^k \cap \underline{\underline{S}}_\pi$ is a Lockett class) that

$$\underline{\underline{Y}}^* \subset \underline{\underline{N}}^k \cap \underline{\underline{S}}_\pi .$$

By (3.2),

$$\underline{\underline{N}} \cap \underline{\underline{S}}_\pi \subseteq \underline{\underline{Y}}^* \text{ so for some } i > 0 ,$$

$$\underline{\underline{N}}^i \cap \underline{\underline{S}}_\pi \subseteq \underline{\underline{Y}}^* \text{ but } \underline{\underline{N}}^{i+1} \cap \underline{\underline{S}}_\pi \not\subseteq \underline{\underline{Y}}^* .$$

It then follows from (3.3) that there is a group $G \in \left(\underline{\underline{N}}^{i+1} \cap \underline{\underline{S}}_\pi\right) \backslash \underline{\underline{Y}}^*$ with the property that for some prime q :

$H = G^{\underline{\underline{N}}}$ is complemented by a cyclic subgroup of order q .

First note that $H \in \underline{\underline{Y}}$. For, since $\underline{\underline{N}}^i \cap \underline{\underline{S}}_\pi \subseteq \underline{\underline{Y}}^*$ it follows from

$$\left(\underline{\underline{N}}^{i} \cap \underline{\underline{S}}_{\pi}\right)_{*} \subseteq \underline{\underline{Y}}^{*}_{*} = \underline{\underline{Y}}_{*} \subseteq \underline{\underline{Y}}$$

that

$$G/G_{\underline{\underline{Y}}} \in Q\{G/G_{(\underline{\underline{N}}^{i} \cap \underline{\underline{S}}_{\pi})_{*}}\} \subseteq \underline{\underline{N}} \quad \text{so} \quad H = G^{\underline{\underline{N}}} \leq G_{\underline{\underline{Y}}} \in \underline{\underline{Y}} .$$

We stress a consequence of the fact that $G \notin \underline{\underline{Y}}^{*}$: whenever a direct power G^{n} of G is embedded normally into a group X a $\underline{\underline{Y}}$-subgroup of X can intersect G^{n} in H^{n} but in nothing larger.

Let $q, q_1, \ldots, q_{k-i}$ be alternately different primes from π . Construct groups G_j, H_j as follows:

$$G_0 = G \ , \ H_0 = H \ ;$$

$$G_j = G_{j-1}^{m(j)} \text{ wr } C_{q_j} \ , \ H_j = H_{j-1}^{m(j)} \text{ wr } C_{q_j} \ , \ 1 \leq j \leq k-i \ ,$$

where $m(j)$ is chosen as in (3.2) corresponding to H_{j-1} and q_j . Note that G_{k-i} has a normal series with abelian factors

$$N_0 < N_1 < \ldots < N_{k-i} = G_{k-i}$$

where N_j is isomorphic to a direct power $G_j^{r(j)}$ of G_j .

Now by (3.2), $H_0, H_1, \ldots, H_{k-i} \in \underline{\underline{Y}}$. Also by the remark in the penultimate paragraph H_j^{r} is a maximal $\underline{\underline{Y}}$-subgroup of G_j^{r} for every positive integer r . Indeed since

$$H_{k-i} \cap N_j = H_j^{r(j)}$$

it follows from [3, Korollar] that H_{k-i} is a $\underline{\underline{Y}}$-injector of G_{k-i} . However $(G_{k-i})_{\underline{\underline{N}}^{k} \cap \underline{\underline{S}}_{\pi}}$ is contained in every $\underline{\underline{N}}^{k} \cap \underline{\underline{S}}_{\pi}$-injector of G_{k-i} and, since $\underline{\underline{Y}} \ll \underline{\underline{N}}^{k} \cap \underline{\underline{S}}_{\pi}$, H_{k-i} is in at least one. But then an $\underline{\underline{N}}^{k} \cap \underline{\underline{S}}_{\pi}$-injector of G_{k-i} contains

$$(G_{k-i})_{\underline{\underline{N}}^{k} \cap \underline{\underline{S}}_{\pi}} \cdot H_{k-i} = G_{k-i}$$

which means $G_{k-i} \in \underline{\underline{N}}^{k}$, a contradiction. The proof is therefore complete.

4. Some general results

In this section we collect together a number of general results, some presumably well-known, others apparently new. Those which are not obvious are established by *ad hoc* methods: if there is any underlying theme for these, it might be described as random attempts to isolate the properties of $\underline{\underline{S}}_\pi$ that give strong containment by general dispensation. We also record some of the questions we have been unable to settle.

The following lemma is well known and trivial.

(4.1) LEMMA. *If* $\underline{\underline{X}}, \underline{\underline{Y}}, \underline{\underline{Z}}$ *are Fitting classes with* $\underline{\underline{X}} << \underline{\underline{Y}}$ *and* $\underline{\underline{Y}} << \underline{\underline{Z}}$ *, then* $\underline{\underline{X}} << \underline{\underline{Z}}$.

(4.2) COROLLARY. *If* $\underline{\underline{X}} << \underline{\underline{Y}}$ *, then* $\underline{\underline{X}} << \underline{\underline{Y}}^*$. *If* $\underline{\underline{X}}^* << \underline{\underline{Y}}$ *, then* $\underline{\underline{X}} << \underline{\underline{Y}}$.

Again, proofs are trivial: the reader is referred to Lockett [5] for definitions and basic properties of the star operation. This result suggests two questions we have not been able to answer.

(4.3) QUESTION. (a) If $\underline{\underline{X}} << \underline{\underline{Y}}^*$, is $\underline{\underline{X}}^* << \underline{\underline{Y}}^*$?

(b) If $\underline{\underline{X}}^* \subseteq \underline{\underline{Y}}$, and $\underline{\underline{X}}^* << \underline{\underline{Y}}^*$, is $\underline{\underline{X}}^* << \underline{\underline{Y}}$?

The next result has as a corollary a method of "localizing" for establishing strong containment.

(4.4) THEOREM. *Let* $\underline{\underline{X}}$ *be a Fitting class,* $\underline{\underline{Y}}$ *an S-closed Fitting class, and suppose* $\underline{\underline{X}} \cap \underline{\underline{Y}} << \underline{\underline{Y}}$. *If* G *is a group,* V *a* $\underline{\underline{Y}}$*-injector of* G *,* U *an* $\underline{\underline{X}}$*-injector of* V *, then* U *is an* $\underline{\underline{X}} \cap \underline{\underline{Y}}$*-injector of* G .

Proof. The proof is by induction on $|G|$, the result being clearly true if $|G| = 1$. Hence suppose the result has been proved for all groups of order less than $|G|$. Let V be a $\underline{\underline{Y}}$-injector of G , U an $\underline{\underline{X}}$-injector of V . Since $\underline{\underline{X}} \cap \underline{\underline{Y}} << \underline{\underline{Y}}$, V contains an $\underline{\underline{X}} \cap \underline{\underline{Y}}$-injector of G , W say.

Now let N be a normal subgroup of G such that $G/N \in \underline{\underline{N}}$. Then $W \cap N$ is an $\underline{\underline{X}} \cap \underline{\underline{Y}}$-injector of N , and $U \cap N$ is an $\underline{\underline{X}}$-injector of $V \cap N$, a $\underline{\underline{Y}}$-injector, and hence by our induction hypothesis is an $\underline{\underline{X}} \cap \underline{\underline{Y}}$-injector of N . Thus we may assume $U \cap N = W \cap N$.

Finally, note that W is a maximal $\underline{\underline{X}}$-subgroup of V : for $W \in \underline{\underline{X}}$, and if $W < T \leq V$ with $T \in \underline{\underline{X}}$, then $T \in \underline{\underline{X}} \cap \underline{\underline{Y}}$ also (since $\underline{\underline{Y}}$ is subgroup closed), a contradiction. Hence we may invoke (2.2) to show that U, W are conjugate, and the theorem is proved.

(4.5) COROLLARY. *Let* $\underline{\underline{X}}, \underline{\underline{Y}}$ *be Fitting classes, and let* π_λ , $\lambda \in \Lambda$ *, be sets of primes such that* $\underline{\underline{S}} = \bigcup_{\lambda \in \Lambda} \underline{\underline{S}}_{\pi_\lambda}$. *Then* $\underline{\underline{X}} << \underline{\underline{Y}}$ *if and only if* $\underline{\underline{X}} \cap \underline{\underline{S}}_{\pi_\lambda} << \underline{\underline{Y}} \cap \underline{\underline{S}}_{\pi_\lambda}$ *for all* $\lambda \in \Lambda$.

Proof. Suppose $\underline{\underline{X}} \ll \underline{\underline{Y}}$, π a set of primes, and G a group. If U is an $\underline{\underline{X}} \cap \underline{\underline{S}}_\pi$-injector of G , then $U \leq H$, a Hall π-subgroup of G , and by Theorem 4.4, U is an $\underline{\underline{X}}$-injector of H . But then $U \leq V$, for some $\underline{\underline{Y}}$-injector V of H , and again by Theorem 4.4, V is a $\underline{\underline{Y}} \cap \underline{\underline{S}}_\pi$-injector of G , giving $\underline{\underline{X}} \cap \underline{\underline{S}}_\pi \ll \underline{\underline{Y}} \cap \underline{\underline{S}}_\pi$. The other direction is easy.

For our next result, we need a special case of Theorem 3.2 of Lockett [5], in which he calculates an injector of a product of Fitting classes (recall that if $\underline{\underline{U}}$, $\underline{\underline{V}}$ are Fitting classes, their product is defined by $\underline{\underline{UV}} = \{G : G/G_{\underline{\underline{U}}} \in \underline{\underline{V}}\}$, and is itself a Fitting class). If π is the smallest set of primes such that the Fitting class $\underline{\underline{U}}$ is contained in $\underline{\underline{S}}_\pi$, we call π the characteristic of $\underline{\underline{U}}$. The following lemma is easily deduced as a special case of Lockett's theorem.

(4.6) LEMMA. *Let $\underline{\underline{X}}$ be a Fitting class of characteristic π , and let $\underline{\underline{Y}}$ be a Fitting class with $\underline{\underline{Y}} \subseteq \underline{\underline{S}}_{\pi'}$. If U is a $\underline{\underline{YX}}$-injector of G , then $U = VW$, $V \triangleleft U$, $V \cap W = 1$, where W is an $\underline{\underline{X}}$-injector of G , and V is a $\underline{\underline{Y}}$-injector of $O_{\pi'}(G)$.*

(4.7) THEOREM. *Let $\underline{\underline{X}}$ be a Fitting class of characteristic π , and suppose $\underline{\underline{X}} \ll \underline{\underline{Y}}$. Then $\underline{\underline{X}} \ll (\underline{\underline{S}}_{\pi'} \cap \underline{\underline{Y}})\underline{\underline{X}} \ll \underline{\underline{Y}}$.*

Proof. It follows immediately from Lemma 4.6 that $\underline{\underline{X}} \ll (\underline{\underline{S}}_{\pi'} \cap \underline{\underline{Y}})\underline{\underline{X}}$. Let U be an $(\underline{\underline{S}}_{\pi'} \cap \underline{\underline{Y}})\underline{\underline{X}}$-injector: then $U = VW$, where $V \triangleleft U$, $V \cap W = 1$, W is an $\underline{\underline{X}}$-injector of G , and V is an $(\underline{\underline{S}}_{\pi'} \cap \underline{\underline{Y}})$-injector of $O_{\pi'}(G)$. But then V is a $\underline{\underline{Y}}$-injector of $O_{\pi'}(G)$, and hence there is a $\underline{\underline{Y}}$-injector A of G such that $A \cap O_{\pi'}(G) = V$, and so $V \triangleleft A$. Since $\underline{\underline{X}} \ll \underline{\underline{Y}}$, we have an $\underline{\underline{X}}$-injector of G , T say, contained in A . Now both W, T are in $N_G(V)$, and so there is a $g \in N_G(V)$ such that $T^g = W$, giving $U = VW = V.T^g = (VT)^g \leq A^g$. But A^g is a $\underline{\underline{Y}}$-injector and hence the result is proved.

This theorem is rather surprising: especially the fact that $\underline{\underline{X}} \ll \underline{\underline{Y}}$ guarantees that $(\underline{\underline{Y}} \cap \underline{\underline{S}}_{\pi'})\underline{\underline{X}}$ is contained in $\underline{\underline{Y}}$. One useful aspect is that in trying to prove $\underline{\underline{X}} \ll \underline{\underline{Y}}$, we may assume, by replacing $\underline{\underline{X}}$ by $(\underline{\underline{Y}} \cap \underline{\underline{S}}_{\pi'})\underline{\underline{X}}$ if necessary, that $\underline{\underline{X}}$ and $\underline{\underline{Y}}$ have the same characteristic.

(4.8) LEMMA. *Let $\underline{\underline{X}}$ be a Fitting class and α a set of primes. Then $\underline{\underline{S}}_\alpha \ll \underline{\underline{XS}}_\alpha$.*

Proof. Follows immediately from Theorem 3.2 of Lockett [5].

Note as a consequence of (4.8) we have that if p, q are distinct primes, then

$\underline{\underline{S}}_p \ll \underline{\underline{S}}_q\underline{\underline{S}}_p$: however it is easy to see that $\underline{\underline{S}}_q$ is not strongly contained in $\underline{\underline{S}}_q\underline{\underline{S}}_p$, for $(C_q \text{ wr } C_p) \text{ wr } C_q$ is easily seen to be a group in which no $\underline{\underline{S}}_q$-injector is contained in the $\underline{\underline{S}}_q\underline{\underline{S}}_p$-injector. The idea behind this example can be put in the following general form.

(4.9) THEOREM. *Suppose that* $\underline{\underline{X}}$ *and* $\underline{\underline{Y}}$ *are Fitting classes, and there exists a prime* p *and a group* G *such that*

(i) $G \in \underline{\underline{Y}}$,

(ii) $G_{\underline{\underline{X}}} = G_{\underline{\underline{X}}*}$ *(=* H *say)* ,

(iii) $H \text{ wr } C_p \in \underline{\underline{X}}$,

(iv) $H \neq G$, $G/H \in \underline{\underline{N}}$,

(v) $G \text{ wr } C_p \notin \underline{\underline{Y}}$.

Then $\underline{\underline{X}}$ *is not strongly contained in* $\underline{\underline{Y}}$.

Proof. Consider a $\underline{\underline{Y}}$-injector of $G \text{ wr } C_p$: it is clearly $\prod_{x \in C_p} G^x$. Now consider an $\underline{\underline{X}}$-injector of $G \text{ wr } C_p$; U say. Then $U \cap \prod G^x \geq \prod H^x$, and cannot be any larger or $G_{\underline{\underline{X}}*}$ would be larger than H . Hence, since $\prod H_x . C_p \cong H \text{ wr } C_p \in \underline{\underline{X}}$, it is not difficult to show that U is $\prod H_x . C_p$ or a conjugate. Thus no $\underline{\underline{X}}$-injector is contained in a $\underline{\underline{Y}}$-injector, and the theorem is proved.

This result can be useful where the situation of (3.3) holds, or where we are dealing with primitive saturated formations. The following result demonstrates this: its proof is by an easy induction on nilpotent length.

(4.10) LEMMA. *Let* $\underline{\underline{X}}$ *be a primitive saturated formation, and* $G \in \underline{\underline{X}}$ *such that* $G/G^{\underline{\underline{N}}} \in \underline{\underline{S}}_p$, p *a prime. Then if* H *is any group such that* $H^{\underline{\underline{N}}} \cong G^{\underline{\underline{N}}}$, *and* $H/H^{\underline{\underline{N}}} \in \underline{\underline{S}}_p$, $H \in \underline{\underline{X}}$.

5. Strong containment in Fitting length two and three

Consider $\underline{\underline{X}} \ll \underline{\underline{Y}}$, $\underline{\underline{Y}} \subseteq \underline{\underline{N}}^2$. Suppose first that $\text{charc } \underline{\underline{X}} = \text{charc } \underline{\underline{Y}}$, and that $\underline{\underline{Y}}$ is a primitive saturated formation. If $\underline{\underline{X}}$ is not strongly normal in $\underline{\underline{Y}}$, then there is a group $G \in \underline{\underline{Y}} \backslash \underline{\underline{X}}*$ of least order. It must be a *non-nilpotent* group from $\underline{\underline{S}}_p\underline{\underline{S}}_q$ for some primes p, q . In particular

$$\underline{\underline{S}}_p\underline{\underline{S}}_q \not\subseteq \underline{\underline{X}}^*$$

and $\underline{\underline{S}}_p\underline{\underline{S}}_q \subseteq \underline{\underline{Y}}$, since $\underline{\underline{Y}}$ is a primitive saturated formation. By (3.3) there exists $H \in \underline{\underline{S}}_p\underline{\underline{S}}_q \backslash \underline{\underline{X}}^*$ such that

$$H^{\underline{\underline{N}}} \in \underline{\underline{S}}_p \ , \quad H = H^{\underline{\underline{N}}}.B \ , \quad |B| = q \ , \quad H_{\underline{\underline{X}}} = H_{\underline{\underline{X}}^*} = H^{\underline{\underline{N}}} \ .$$

Since it has the wrong nilpotent length H wr $C_p \notin \underline{\underline{Y}}$, and so (4.9) gives a contradiction. Hence $\underline{\underline{X}}$ is strongly normal in $\underline{\underline{Y}}$.

Suppose now π = charc $\underline{\underline{X}} \neq$ charc $\underline{\underline{Y}}$. Then by (4.7),

$$(\underline{\underline{Y}} \cap \underline{\underline{S}}_{\pi'})\underline{\underline{X}} << \underline{\underline{Y}} \ ,$$

and if $\underline{\underline{X}}$ is not nilpotent $(\underline{\underline{Y}} \cap \underline{\underline{S}}_{\pi'})\underline{\underline{X}}$ has nilpotent length 3, a contradiction. Thus, as above $((\underline{\underline{Y}} \cap \underline{\underline{S}}_{\pi'})\underline{\underline{X}})^* = \underline{\underline{Y}}$. But $(\underline{\underline{Y}} \cap \underline{\underline{S}}_{\pi'})\underline{\underline{X}}$ is a Lockett class and so

$$(\underline{\underline{Y}} \cap \underline{\underline{S}}_{\pi'})\underline{\underline{X}} = \underline{\underline{Y}} \ .$$

We have established the following result.

(5.1) THEOREM. *Suppose* $\underline{\underline{X}} << \underline{\underline{Y}} \subseteq \underline{\underline{N}}^2$, *where* $\underline{\underline{Y}}$ *is a primitive saturated formation. Then either*

(i) charc $\underline{\underline{X}}$ = charc $\underline{\underline{Y}}$ *and* $\underline{\underline{X}}$ *is strongly normal in* $\underline{\underline{Y}}$, *or*

(ii) π = charc $\underline{\underline{X}} \neq$ charc $\underline{\underline{Y}}$, *and* $\underline{\underline{Y}} = (\underline{\underline{Y}} \cap \underline{\underline{S}}_{\pi'})\underline{\underline{X}}$.

Though we can say quite a deal about the nilpotent length three case, especially for primitive saturated formations, and though a general pattern seems to be hiding in the shadows, the results are too diffuse to record them all here. We content ourselves with a couple of results which are the best illustrations. Proofs are not given in detail.

Suppose that $\underline{\underline{X}} << \underline{\underline{Y}} \subseteq \underline{\underline{N}}^3$ when π = charc $\underline{\underline{X}}$ = charc $\underline{\underline{Y}}$ and suppose further that $\underline{\underline{X}}^*$ and $\underline{\underline{Y}}$ are primitive saturated formations.

(i) If $G \in \underline{\underline{Y}} \backslash \underline{\underline{X}}^*$ has minimal order then G is metanilpotent. Hence

(ii) if $\underline{\underline{X}}^* \cap \underline{\underline{N}}^2 = \underline{\underline{Y}} \cap \underline{\underline{N}}^2$ then $\underline{\underline{X}}^* = \underline{\underline{Y}}$ (so $\underline{\underline{X}}$ is strongly normal in $\underline{\underline{Y}}$).

However

(iii) if $\underline{\underline{X}}^* \cap \underline{\underline{N}}^2 < \underline{\underline{Y}} \cap \underline{\underline{N}}^2$ so that $\underline{\underline{S}}_p\underline{\underline{S}}_q \subseteq \underline{\underline{Y}}$ but $\underline{\underline{S}}_p\underline{\underline{S}}_q \not\subseteq \underline{\underline{X}}^*$ for some primes p, q then either

(a) $\underline{\underline{Y}} \cap \underline{\underline{S}}_{\{p,q\}} = \underline{\underline{S}}_p\underline{\underline{S}}_q\underline{\underline{S}}_p$ and $\underline{\underline{X}}^* \cap \underline{\underline{S}}_{\{p,q\}} = \underline{\underline{S}}_q\underline{\underline{S}}_p$ or

(b) $\underline{\underline{Y}} \cap \underline{\underline{S}}_{\{p,q\}} = \underline{\underline{N}}^3_{\{p,q\}}$ and $\underline{\underline{X}}^* \cap \underline{\underline{S}}_{\{p,q\}} = \underline{\underline{N}}_{\{p,q\}}$.

References

[1] R.A. Bryce and John Cossey, "Metanilpotent Fitting classes", *J. Austral. Math. Soc.* 17 (1974), 285-304. Zb1M292#20016.

[2] R.A. Bryce and John Cossey, "A problem in the theory of normal Fitting classes", *Math. Z.* 141 (1975), 99-110. Zb1M283#20016

[3] B. Fischer, W. Gaschütz und B. Hartley, "Injektoren endlicher auflösbarer Gruppen", *Math. Z.* 102 (1967), 337-339. MR36#6504.

[4] B. Hartley, "On Fischer's dualization of formation theory", *Proc. London Math. Soc.* (3) 19 (1969), 193-207. MR39#5696.

[5] F. Peter Lockett, "On the theory of Fitting classes of finite soluble groups", *Math. Z.* 131 (1973), 103-115. MR47#6847.

[6] F. Peter Lockett, "The Fitting class $\underline{\underline{F}}^*$ ", *Math. Z.* 137 (1974), 131-136. Zb1M286#20017.

[7] A.R. Makan, "Normal Fitting classes and the Lockett ordering", *Math. Z.* 142 (1975), 221-228. Zb1M284#20026.

Department of Pure Mathematics,
School of General Studies,
Australian National University,
Canberra, ACT, Australia.

PROC. MINICONF. THEORY OF GROUPS
CANBERRA 1975, 17-50. 20G15, 20C15

NONABELIAN SUBGROUPS OF PRIME-POWER ORDER OF CLASSICAL GROUPS OF THE SAME PRIME DEGREE

S.B. Conlon

Let p be a prime and F a field, not of characteristic p. Finite nonabelian p-subgroups of the following groups of degree p with coefficients in F are studied: GL, SL, PGL, PSL, U, SU, PU, PSU. The isomorphism and conjugacy classes of p-groups are determined together with their normalisers and p-overgroups.

PART I. GENERALITIES

1. Introduction

Let p be a prime and let F be a field, not of characteristic p. In Conlon (1975), the nonabelian finite p-subgroups of $GL(p, F)$ were discussed under the assumption that F contained all p-power roots of 1. Here we drop the last assumption and extend the work to SL, PGL, PSL, U, SU, PU and PSU. The whole discussion is possible because of the nonmodular situation and the precise degree p involved. $Sp(2, F)$ and $O(2, F)$ are not discussed as the former is related to $SL(2, F)$ and the latter only involves dihedral groups.

The treatment depends on Conlon (1975) and all unexplained notation will be found therein. The isomorphism classes occurring in each case, together with a realisation of a group in each conjugacy class will be given. The normalisers are described and put down explicitly. A *p-overgroup* of a p-group P is a p-group in which P is maximal. These are put down when unique.

If $- : F \to F$, $\phi \mapsto \overline{\phi}$ is an involutary automorphism of F , we define

$$U(p, F) = \{g \in GL(p, F) \mid g\overline{g}^t = I\} ,$$

t denoting the transpose and I the unit matrix.

Part II considers the case when p is an odd prime and follows the notation of Conlon (1975) exactly. Part III discusses $p = 2$ and although some results can be described in the notation of Conlon (1975) and are repetitions of portions of II, a change of notation is required for the last parts and so the whole of III is done in this notation to add cohesion and to allow comparisons. Part I contains some general properties. Because of length, proofs are not given.

As set out in Conlon (1975), the different isomorphism types of nonabelian finite p-groups are described as certain subgroups P_{klm} of $Z_{p^\infty} \text{ wr } Z_p$; this last is regarded as embedded in $Z_{p^\infty} \text{ wr } \Sigma_p$ which in turn is regarded as embedded in $GL(p, F^*)$, where F^* is the algebraic closure of F , Σ_p is given by permutation matrices and $Z_{p^\infty} \times \ldots^{(p)} \ldots \times Z_{p^\infty}$ is given by diagonal matrices of p power roots of 1 . k is the class of P_{klm} , p^l is the order of the center, the group has order p^{k+l} and $m \in (0, 1, \ldots, p)$ for $k > 2$ and $m \in (0, p)$ for $k = 2$. When it is necessary to designate the prime p , it is included as a fourth suffix, P_{klmp} .

$Z(p, F) = F^{\times}I$ is the center (scalar matrices) of $GL(p, F)$. The superscript "$\times$" means "the nonzero elements of". If $H \le GL(p, F)$, we write

$$PH = F^{\times}H/F^{\times}I \simeq H/\left(H \cap F^{\times}I\right) .$$

Write $u = u(F)$ for the largest integer such that F contains a primitive p^uth root of 1 . We allow $u(F) = \infty$.

The author wishes to thank the Mathematics Department, University of Illinois at Urbana, for having appointed him a Visiting Research Associate in 1974/1975 during which time this work was done.

2. Miscellaneous properties

(2.1) If $H \le GL(p, F)$, any nonabelian finite p-subgroup of H with class greater than 2 lies in a unique maximal p-subgroup (perhaps infinite) of H . Any nonabelian finite p-subgroup of PH lies in a unique maximal p-subgroup (perhaps infinite) of PH and can have at most one p-overgroup in PH .

(2.2) If $H \le SL(p, F)$, then any nonabelian finite p-subgroup of H of class greater than 2 can have at most one p-overgroup in H .

(2.3) Isomorphic maximal abelian p-subgroups of $GL(p, F)$ are conjugate.

(2.4) Isomorphic maximal p-subgroups of $GL(p, F)$ are conjugate.

(2.5) Isomorphic finite nonabelian p-subgroups of $GL(p, F)$ are conjugate (use Deuring's Theorem and 4.2 of Conlon (1975)).

(2.6) We have the following diagram of inclusions *up to isomorphism* $(0 < m < p)$:

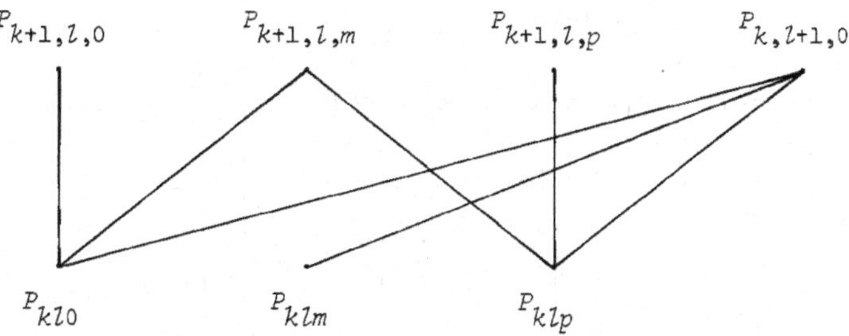

(2.7) Suppose $|F| < \infty$ and $H \leq GL(p, F)$. Then the only nonabelian p-subgroup of H which can be a defect group of a p-block of H is the Sylow p-subgroup S of H . In this case the blocks of H with defect group S are in one-to-one correspondence with the linear characters of $(F^{\times}I \cap H)_{p'}$.

(2.8) Suppose $|F| < \infty$ and $H \leq GL(p, F)$. Then the only p-block of PH which can have a nonabelian defect group is the principal block.

PART II. CASE OF p ODD

3. $GL(p, F)$, p odd

(3.1) We assume p is odd throughout the whole of Part II. To have any non-abelian p-subgroups in $GL(p, F)$, $u = u(F)$ must be greater than 0 . The Sylow p-subgroup of $GL(p, F)$ is then isomorphic to $Z_{p^u} \text{ wr } Z_p$. If $u = \infty$, all the P_{klm} occur in GL . If $u < \infty$, we get all subgroups of

$$Z_{p^u} \text{ wr } Z_p \approx P_{u(p-1)+1,u,m(u)} ,$$

given by the inclusion diagram (2.6), where $m(u) = 1$ or $p - 1$ and $m(u) \equiv (-1)^{u-1} \bmod p$. Thus when $u = u(F) < \infty$, GL contains (p. 21)

Table 1

Group	OSA	Gens	p-overgroups
P_{2l0} $(1 \le l < u,\ u(p-1) > 2)$	$\mathrm{SL}(2,\ \mathsf{F}_p)$	$y_3,\ \omega$	$P_{2,l+1,0}$, $p+1$ groups iso[2] to $P_{3,l,m}$ for each $m \in (0,\ \ldots,\ p-1)$
P_{2u0} $(2 < u(p-1) < \infty)$	$\mathrm{SL}(2,\ \mathsf{F}_p)$	$y_3,\ \omega$	$p+1$ groups iso. to P_{3u0} [2]
P_{2103} $(u = 1)$	$\mathrm{SL}(2,\ \mathsf{F}_3)$	$y_3^{(1)},\ \omega$	4 groups iso. to P_{3113} [2]
P_{kl0} $(2 < k < u(p-1),\ 1 \le l < u)$	metacyclic[1]	$y_{k+1},\ \kappa(n)$	$P_{k,l+1,0}$, $P_{k+1,l,m}$ $(0 \le m < p)$
P_{ku0} $(2 < k < u(p-1) < \infty)$	metacyclic[1]	$y_{k+1},\ \kappa(n)$	$P_{k+1,u,0}$
$P_{u(p-1),l,0}$ $(1 \le l < u < \infty,\ u(p-1) > 2)$	metacyclic[1]	$y_{u(p-1)+1}^{(m(u))},\ \kappa(n)$	$P_{u(p-1),l+1,0}$
$P_{u(p-1),u,0}$ $(1 \le u < \infty,\ u(p-1) > 2)$	metacyclic[1]	$y_{u(p-1)+1}^{(m(u))},\ \kappa(n)$	$P_{u(p-1)+1,u,m(u)}$
P_{klp} $(2 \le k < u(p-1),\ 1 \le l < u)$	Z_p	y_{k+1}	$P_{k,l+1,0}$, $P_{k+1,l,p}$ and $p-1$ groups[3] iso. to $P_{k+1,l,m}$ $(0 < m < p)$
P_{kup} $(2 \le k < u(p-1) < \infty)$	Z_p	y_{k+1}	$P_{k+1,u,p}$
$P_{u(p-1),l,p}$ $(1 \le l < u < \infty)$	Z_p	$y_{u(p-1)+1}^{(m(u))}$	$P_{u(p-1),l+1,0}$
$P_{u(p-1),u,p}$ $(1 \le u < \infty)$	Z_p	$y_{m(p-1)+1}^{(m(u))}$	$P_{u(p-1)+1,u,m(u)}$
P_{klm} $(3 \le k < u(p-1)+1,\ 0 < m < p,\ 1 \le l < u)$	$Z_{(k-1,p-1)}$	$\kappa(n)$	$P_{k,l+1,0}$
$P_{u(p-1)+1,u,m(u)}$ $(1 \le u < \infty)$	Z_{p-1}	$\kappa(n)$	-

[1] OSA is metacyclic of order $p(p-1)$ and the centraliser of the subgroup of order p in OSA has order $(k,\ p-1)p$.

[2] These overgroups come from the $p + 1$ Sylow p-subgroups of $\mathrm{SL}(2,\ \mathsf{F}_p)$.

[3] These $p - 1$ groups are precisely $\left(P_{k+1,l,m}\right)^{\left(y_{k+2}\right)^{m'}}$, when $k < u(p-1) - 1$ and $\left(P_{u(p-1),l,m}\right)^{\left(y_{u(p-1)+1}^{(m(u))}\right)^{m'}}$, when $k = u(p-1) - 1$ and where $0 < m,\ m' < p$ and $mm' \equiv 1 \bmod p$.

$$P_{kl0} \text{ and } P_{klp} \quad (2 \le k \le u(p-1),\ 1 \le l \le u) ,$$

$$P_{klm} \quad (3 \le k \le u(p-1),\ 1 \le l < u,\ 0 < m < p)$$

and

$$P_{u(p-1)+1,u,m(u)} .$$

We write $\mathrm{OSA}(P_{klm}) = N_{\mathrm{GL}}(P_{klm})/F^{\times}P_{klm}$ (Outer Similarity Automorphism group - see Conlon (1975)). In Table 1 the column "gens" gives elements of GL which give generators in OSA .

For the definitions of y_k, $y_{u(p-1)+1}^{(m(u))}$, ω and $\kappa(n)$, see Conlon (1975).

4. SL(p, F) , p odd

The allowable groups in SL(p, F) are P_{k10} for $2 \le k \le u(p-1)$, where $u = u(F) > 0$. A conjugacy class of subgroups of SL isomorphic to P_{k10} is determined by the determinant of a matrix from GL which conjugates the standard P_{k10} into an element of the conjugacy class of SL under question. The residue class of this determinant value in $F^{\times}$ modulo determinants of elements in $N_{\mathrm{GL}}(P_{k10})$ is the invariant of the orbit. The set of conjugacy classes corresponds to a quotient of $F^{\times}$, called the *conjugacy set*. p-overgroups are unique and the conjugacy class of the overgroup is given by passing to the image quotient of $F^{\times}$ corresponding to the conjugacy set of the larger group.

Write $\mathrm{OSA}(P_{k10}) = N_{\mathrm{SL}}(P_{k10})/P_{k10}$.

If $u = \infty$, a maximal p-subgroup is

$$P_{\infty 10} = \lim_{k\to} P_{k10} ,$$

and is generated by all diagonal matrices $\mathrm{diag}(\alpha_1, \ldots, \alpha_p)$, where the α_i are p power roots of 1 and $\prod \alpha_i = 1$, and the element x (see Conlon (1975)). $P_{\infty 10}$ has similarity automorphisms $\kappa(n)$ and $\mathrm{OSA}(P_{\infty 10}) \approx Z_{p-1}$. The conjugacy set of $P_{\infty 10}$ is $F^{\times}/F^{\times p}$, where $F^{\times p}$ is the subgroup of $F^{\times}$ of all pth powers of elements of $F^{\times}$.

We write F_p for the subgroup of $F^{\times}$ consisting of p power roots of 1 . When $u < \infty$ we have the following.

Table 2

Groups		OSA	Gens	p-overgroups	Conjugacy set
P_{210}	$(u(p-1) > 2)$	$SL(2, F_p)$	y_3, ω [1]	$p+1$ groups[2] iso. to P_{310}	$F^\times/F^{\times p}$
P_{2103}	$(u = 1)$	Q_8 [3]	ω, $\omega^{y_3^{(1)}}$	-	$F^\times/\langle F^{\times 3}, F_3\rangle$
P_{k10}	$(2 < k < u(p-1))$	metacyclic[4]	y_{k+1}, $\kappa(n)$	$P_{k+1,1,0}$	$F^\times/F^{\times p}$
$P_{u(p-1),1,0}$		Z_{p-1}	$\kappa(n)$	-	$F^\times/\langle F^{\times p}, F_p\rangle$

[1] ω is given by similarity action of the matrix $(t(\zeta))^{-1}$ (see proof of 4.1 in Conlon (1975)). $t(\zeta)$ is the $p \times p$ van der Monde matrix $(\zeta^{(i-1)(j-1)})$, where ζ is a primitive pth root of 1. $(t(\zeta))^{-1} = (1/p).t(\zeta^{-1})$ and $|t(\zeta)| = \pm(1-\zeta)^p \dots \left(1 - \zeta^{\frac{p-1}{2}}\right)^p$, and so $|t(\zeta)|$ is a pth power in $F^\times$.

[2] These groups correspond to the $p+1$ Sylow p-subgroups of $SL(2, F_p)$.

[3] Q_8 $(\approx P_{2122})$ is the normal closure in $SL(2, F_3)$ of the generator "ω".

[4] OSA is metacyclic of order $p(p-1)$ and the centraliser in OSA of the subgroup of order p has order $(k, p-1)p$.

5. PGL(p, F), p odd

The only finite nonabelian p-subgroups of $PGL(p, F)$ are P_{k10} $(2 \le k \le u(p-1))$ (note that $P_{klm}/Z(P_{klm}) \approx P_{k-1,1,0}$ for $k \ge 3$). One works with preimages back in GL. Such a preimage of P_{k10} is conjugate to a subgroup $Q_{k+1}(\rho) \le GL$, where $\rho \in F^\times$.

Here

$$Q_k(\rho) = F^\times . \langle x(\rho), y_1, \dots, y_k\rangle, \text{ if } k < u(p-1) + 1$$

and

$$Q_k(\rho) = F^\times . \left\langle x(\rho), y_1, \dots, y_{k-1}, y_k^{(m(u))}\right\rangle,$$

$\left\{y_k^{(m(u))} \in P_{u(p-1)+1,u,m(u)}\right\}$ if $k = u(p-1) + 1 < \infty$, and where $x(\rho)$ is the matrix with ρ in the top right corner and with 1 in each subdiagonal position. (Thus $x = x(1)$.) The $Q_k(\rho)$ $(k > 2)$ have one abelian maximal subgroup containing $F^\times I$

and conjugacy classes are determined by determinants of elements of $Q_k(\rho)$ outside this subgroup. Thus we obtain conjugacy sets $\{1\} \cup \pi(F^\times/F^{\times p})$ for P_{k10} in PGL if $k < u(p-1)$ and $\{1\} \cup \pi\left(F^\times/\left\langle F^{\times p}, F_p\right\rangle\right)$ if $k = u(p-1) < \infty$. Here π denotes the set of projective points corresponding to rays of the vector space over $\mathbf{F}_p$ within parenthesis; $\{1\}$ corresponds to the zero subspace of this vector space. For $P_{k10} \leq \mathrm{PGL}$, write

$$\mathrm{OSA}(P_{k10}) = N_{\mathrm{PGL}}(P_{k10})/P_{k10} \quad \left(\approx N_{\mathrm{GL}}(Q_{k+1}(\rho))/Q_{k+1}(\rho)\right) .$$

In Table 3 the generators of $\mathrm{OSA}(P_{k10})$ are given by matrices in GL operating on the preimage $Q_{k+1}(\rho)$. p-overgroups are unique when they exist and the conjugacy class of the overgroup is obtained by passing to the quotient in the appropriate set.

Table 3

Groups	Parameter	OSA	Gens	p-overgroups
P_{k10} $(2 \leq k < u(p-1)-1)$	$\rho = 1$	metacyclic[1]	y_{k+2}, $\kappa(n)$	$P_{k+1,1,0}$
P_{k10} $(2 \leq k < u(p-1)-1)$	$\rho \notin F^{\times p}$	Z_p	y_{k+2}	$P_{k+1,1,0}$
$P_{u(p-1)-1,1,0}$	$\rho = 1$	metacyclic[1]	$y_{u(p-1)+1}^{(m(u))}$, $\kappa(n)$	$P_{u(p-1),1,0}$
$P_{u(p-1)-1,1,0}$	$\rho \notin F^{\times p}$	Z_p	$y_{u(p-1)+1}^{(m(u))}$,	$P_{u(p-1),1,0}$
$P_{u(p-1),1,0}$	$\rho = 1$	Z_{p-1}	$\kappa(n)$	-
$P_{u(p-1),1,0}$	$\rho \notin \left\langle F^{\times p}, F_p\right\rangle$	(1)	-	-

[1] OSA is metacyclic of order $p(p-1)$ and the centraliser in OSA of the subgroup of order p has order $(k+1, p-1)p$.

6. $\mathrm{PSL}(p, F)$, p odd

We regard PSL as embedded in PGL. Looking at preimages in GL, as $x(\rho)$ in $Q_k(\rho)$ has determinant ρ and as we seek those $Q_k(\rho) \leq F^\times \cdot \mathrm{SL}$, we have $\rho = 1$, and any two copies of P_{k10} in PSL are conjugate under the action of elements from PGL. A conjugacy class of P_{k10} in PSL is determined by the determinant of the conjugating element in GL which sends the standard copy of $Q_{k+1}(1)$ into the preimage of P_{k10} under question, modulo determinants of elements in $N_{\mathrm{GL}}(Q_{k+1}(1))$. Thus the conjugacy set is a quotient of $F^\times$. Write

$\mathrm{OSA}(P_{k10}) = N_{\mathrm{PSL}}(P_{k10})/P_{k10}$. p-overgroups are unique and conjugacy classes of such are obtained by passing to the quotient in the appropriate conjugacy sets.

Table 4

Group	OSA	Gens	p-overgroups	Conjugacy Set
P_{k10} $(2 \le k < u(p-1)-1)$	metacyclic[1]	y_{k+2}, $\kappa(n)$	$P_{k+1,1,0}$	$F^\times/F^{\times p}$
$P_{u(p-1)-1,1,0}$	Z_{p-1}	$\kappa(n)$	-	$F^\times/\langle F^{\times p}, F_p\rangle$

[1] OSA is metacyclic of order $p(p-1)$ and the centraliser in OSA of the subgroup of order p has order $(k+1, p-1)p$.

7. $\mathrm{U}(p, F)$, p odd

For p odd and $u = u(F) > 0$, we have no nonabelian p-subgroups in $\mathrm{U} = \mathrm{U}(p, F)$ unless the involutary automorphism $- : F \to F$, $\phi \mapsto \overline{\phi}$, carries each element of the group F_p of p power roots of 1 to its inverse.

If G is an absolutely irreducible group of unitary matrices and if $G^g \le \mathrm{U}$ for some $g \in \mathrm{GL}$, then there exists a scalar multiple g' of g with $g' \in \mathrm{U}$. Hence nonabelian finite p-subgroups of U are conjugate iff they are isomorphic. We obtain exactly the same results as in Section 3 for GL .

The only point requiring comment is that the matrix $(t(\zeta))^{-1}$ which provides the similarity transformation ω for P_{2l0} must be altered by a scalar multiple to lie in U . Note that $t(\zeta) \cdot t(\overline{\zeta}) = t(\zeta) \cdot t(\zeta^{-1}) = pI$. If $\lambda = p^{(p+1)/2}\delta^{-1}$, where $\delta = |t(\zeta)|$, then $\lambda^{-1}t(\zeta) \in \mathrm{U}$. (Indeed $\lambda^{-1}t(\zeta) \in \mathrm{SU}$.)

8. $\mathrm{SU}(p, F)$, p odd

Once again to get nonabelian finite p-subgroups in SU , - must invert F_p . However, when this is true, the situation is exactly the same as for $\mathrm{SL}(p, F)$, as in Section 4. The conjugacy classes of the P_{k10} are simply the intersections with SU of those of SL .

Let $N : F^\times \to N(F^\times) \le F^\times$, $\phi \mapsto \phi\overline{\phi}$ be the norm homomorphism. If $g \in \mathrm{U}$, then $\det g \in N^{-1}(1)$. Hence we have the following conjugacy sets when $u = u(F)$:

Table 5

Group	Conjugacy Set
P_{k10} $(2 \le k < u(p-1))$	$N^{-1}(1)F^{\times p}/F^{\times p} \approx N^{-1}(1)/(N^{-1}(1))^p$
P_{k10} $(2 \le k = u(p-1) < \infty)$	$N^{-1}(1)F^{\times p}F_p/F^{\times p}F_p \approx N^{-1}(1)/(N^{-1}(1))^p F_p$

9. PU(p, F) , p odd

We regard PU as embedded in PGL and we must look at which $Q_k(\rho)$ lie in $F^\times$. U . It must be shown that if an isomorphic copy G of $Q_k(\rho')$ lies in $F^\times$. U , then there exists $u \in$ U such that G^u is in standard form $Q_k(\rho)$ for some $\rho \in F^\times$. The element $x(\rho)$ lies in U iff $N(\rho) = 1$. Again we obtain that the conjugacy classes in PU are simply the intersections of conjugacy classes in PGL with PU and other than this the situation is exactly the same as in Section 5. - must continue to invert the elements of F_p . We have the following conjugacy sets for P_{k10}'s in PU .

Table 6

Group	Conjugacy Set
P_{k10} $(2 \le k < u(p-1))$	$\{1\} \cup \pi(N^{-1}(1)F^{\times p}/F^{\times p}) \approx \{1\} \cup \pi(N^{-1}(1)/N^{-1}(1)^p)$
P_{k10} $(k = u(p-1) < \infty)$	$\{1\} \cup \pi\left(N^{-1}(1)F^{\times p}F_p/F^{\times p}F_p\right) \approx \{1\} \cup \pi\left(N^{-1}(1)/N^{-1}(1)^p F_p\right)$

10. PSU(p, F) , p odd

- must invert the elements of F_p to get any finite nonabelian p-subgroups of PSU . We have that if $Q_k(\rho) \le F^\times$. SU , then $\rho = 1$. Then all copies of $Q_k(1)$ are conjugate in $F^\times$. U and the situation is parallel to that in Section 6. The conjugacy classes of PSU are the restrictions of conjugacy classes of PSL to PSU . We have the following conjugacy sets in PU .

Table 7

Group	Conjugacy Set
P_{k10} $(2 \le k < u(p-1))$	$N^{-1}(1)F^{\times p}/F^{\times p} \approx N^{-1}(1)/N^{-1}(1)^p$
P_{k10} $(k = u(p-1) < \infty)$	$N^{-1}(1)F^{\times p}F_p/F^{\times p}F_p \approx N^{-1}(1)/N^{-1}(1)^p F_p$

PART III. $p = 2$

11. GL(2, F)

In the algebraic closure F^* of F, choose a succession of elements α_i such that $\alpha_0 = 1$, $\alpha_1 = -1$ and for $i > 1$ we have $\alpha_i^2 = \alpha_{i-1}$. Hence α_n is a primitive 2^nth root of 1. Write $\alpha_2 = \iota$, a fourth root of 1. For $n \geq 0$ define

$$\kappa_n = \frac{\alpha_n+\alpha_n^{-1}}{2} \quad \left(\text{"}\cos \frac{\pi}{2^{n-1}}\text{"}\right) \quad \text{and} \quad \sigma_n = \frac{\alpha_n-\alpha_n^{-1}}{2\iota} \quad \left(\text{"}\sin \frac{\pi}{2^{n-1}}\text{"}\right).$$

The following "trigonometric" formulae hold:

$$\kappa_n = 2\kappa_{n+1}^2 - 1 = 1 - 2\sigma_{n+1}^2, \quad \sigma_n = 2\sigma_{n+1}\kappa_{n+1}, \quad \sigma_n^2 + \kappa_n^2 = 1,$$

$$\kappa_n \pm \iota\sigma_n = \alpha_n^{\pm 1}, \quad \kappa_0 = 1, \quad \kappa_1 = -1, \quad \kappa_2 = 0, \quad \sigma_0 = 0, \quad \sigma_1 = 0, \quad \sigma_2 = 1.$$

For $n \geq 3$ both κ_n and $\sigma_n \neq 0$.

We write $u = u(F)$ and $v = u(F(\iota))$. Thus $u \geq 1$ and $v \geq 2$. Also $u = v$, unless $u = 1$. If $u = 1$, then we have $2 \leq v \leq \infty$.

LEMMA 11.1. *Suppose* $u = 1$ *and* $n \geq 3$.

(i) If any one of κ_n, σ_n, $\iota\kappa_n$ *and* $\iota\sigma_n$ *lies in* F, *then so do* $\kappa_0, \ldots, \kappa_{n-1}$ *and* $\sigma_0, \ldots, \sigma_{n-1}$.

(ii) $\kappa_n \in F$ *iff* $\sigma_n \in F$. $\iota\kappa_n \in F$ *iff* $\iota\sigma_n \in F$.

(iii) One of κ_n *and* $\iota\kappa_n \in F$ *iff* $n \leq v$.

(iv) If $\iota\kappa_n \in F$, *then* $n = v$ *and so* v *is finite.*

(11.2) As noted in the introduction we depart from the realisations of the P_{klm2} in GL(2, F) as set out in Conlon (1975) and introduce the following matrices in GL(2, F^*):

$$x = \begin{pmatrix} 0 & 1 \\ 1 & 0 \end{pmatrix}, \quad y_k = \begin{pmatrix} \kappa_k & (-1)^k\sigma_k \\ (-1)^{k+1}\sigma_k & \kappa_k \end{pmatrix} \quad \text{and} \quad z_l = \alpha_l I.$$

Set $P_{kl02} = \langle x, y_k, z_l \rangle$. To realise P_{kl12} and P_{kl22} we set $y_k^{(1)} = \alpha_{l+1}y_k$ and $x' = \alpha_{l+1}x$ and then $P_{kl12} = \left\langle x, y_k^{(1)} \right\rangle$ and $P_{kl22} = \langle x', y_k \rangle$. (p. 28)

Table 8

Group	OSA	Gens	2-overgroups
P_{2102} $(u = 2)$	Z_2	$\alpha_3 y_3$	P_{2202}
P_{2102} $(u > 2)$	Z_2	y_3	P_{2202}, P_{3102}, P_{3112}
P_{2202} $(u = 2)$	Σ_3[1]	ω, $\alpha_3 y_3$	3 groups iso. to P_{3212}[2]
P_{2l02} $(u \geq 3,\ 2 \leq l < u)$	Σ_3	ω, y_3	$P_{2,l+1,0,2}$, 3 groups iso.[2] to $P_{3,l,m,2}$ for each $m \in (0, 1)$
P_{2u02} $(3 \leq u < \infty)$	Σ_3	ω, y_3	3 groups iso. to P_{3u02}[2]
P_{kl02} $(3 \leq k < u,\ 1 \leq l < u)$	Z_2	y_{k+1}	$P_{k,l+1,0,2}$, $P_{k+1,l,m,2}$ for $m \in (0, 1)$
P_{ku02} $(3 \leq k < u < \infty)$	Z_2	y_{k+1}	$P_{k+1,u,0,2}$
P_{ul02} $(3 \leq u < \infty,\ 1 \leq l < u)$	Z_2	$\alpha_{u+1} y_{u+1}$	$P_{u,l+1,0,2}$
P_{uu02} $(3 \leq u < \infty)$	Z_2	$\alpha_{u+1} y_{u+1}$	$P_{u+1,u,1,2}$
P_{kl12} $(3 \leq k < u+1,\ 1 \leq l < u)$	(1)	-	$P_{k,l+1,0,2}$
$P_{u+1,u,1,2}$ $(u < \infty)$	(1)	-	-
P_{2122} $(u = 2)$	Σ_3	ω, $\alpha_3 y_3$	P_{2202}
P_{2122} $(u > 2)$	Σ_3	ω, y_3	P_{2202}, 3 groups iso.[2] to P_{31m2} for each $m \in (1, 2)$
P_{kl22} $(\neq P_{2122})$ $(2 \leq k < u,\ 1 \leq l < u)$	Z_2	y_{k+1}	$P_{k,l+1,0,2}$, $P_{k+1,l,2,2}$ and one group iso. to[3] $P_{k+1,l,1,2}$
P_{ku22} $(2 \leq k < u < \infty)$	Z_2	y_{k+1}	$P_{k+1,u,2,2}$
P_{ul22} $(u > 2,\ 1 \leq l < u < \infty)$	Z_2	$\alpha_{u+1} y_{u+1}$	$P_{u,l+1,0,2}$
P_{uu22} $(2 \leq u < \infty)$	Z_2	$\alpha_{u+1} y_{u+1}$	$P_{u+1,u,1,2}$

1 $\Sigma_3 \approx SL(2, F_2)$.

2 The 3 overgroups come from the Sylow 2-subgroups of Σ_3 .

3 More precisely this group is $(P_{k+1,l,1,2})^{y_{k+2}}$ when $k < u - 1$ and $(P_{ul12})^{y^{(1)}_{u+1}}$ when $k = u - 1 < \infty$.

(11.3) In this new notation the matrix giving the similarity transformation ω of P_{2l02} $(l \geq 2)$ is diag$(1, \iota)$. The similarity transformation $\omega : y_2 \to x'$, $x' \to y_2$ of P_{2122} is given by the matrix $\begin{pmatrix} \iota & 1 \\ -1 & -\iota \end{pmatrix}$.

(11.4) If $u > 1$, we have the following P_{klm2}'s in GL$(2, F)$ (see Table 8).

Towards an analysis of the case $u = 1$, we write $s(F)$ (*stufe* or *level*) for the smallest number of terms required to express -1 as a sum of squares in F. If the characteristic of F is odd, then $s(F)$ is 1 or 2. If the characteristic of F is zero, then $s(F)$ is a power of 2 or ∞ (Pfister's Theorem). If F is an algebraic number field then $s(F)$ is 1, 2, 4 or ∞ (Siegel's Theorem).

LEMMA 11.5. GL$(2, F)$ *contains the quaternion group* Q_8 $(\approx P_{2122})$ *iff* $s(F) \leq 2$.

(11.6) Indeed if $u = 1$ and if $s(F) = 2$, there exist $\xi, \eta \in F$ such that $-1 = \xi^2 + \eta^2$. We modify the matrix x' to $x'' = \begin{pmatrix} \xi & \eta \\ \eta & -\xi \end{pmatrix}$ and so $x''^2 = -I$ as before. We realise P_{k122} as $P_{k122} = \langle y_k, x'' \rangle$ $(k \geq 2)$. The similarity automorphism $\omega : y_2 \to x''$, $x'' \to y_2$ of P_{2122} is realised by the matrix $\begin{pmatrix} \xi & \eta+1 \\ \eta-1 & -\xi \end{pmatrix}$ of determinant 2. If $v = u(F(\iota)) < \infty$ and $\iota\kappa_v \in F$, then $(\iota\kappa_v)^2 + (\iota\sigma_v)^2 = -1$ and in x'' we can take $\xi = \iota\kappa_v$ and $\eta = \iota\sigma_v$. Note also that if $v = 2$, then $\sqrt{2} \notin F$ as $1/\sqrt{2} = \kappa_3 \notin F$. If $u = 1$, then nonabelian finite 2-subgroups of GL$(2, F)$ have centers of order 2. If $|F| < \infty$ and $u = 1$, then $\iota\kappa_v \in F$.

(11.7) We obtain the following finite nonabelian 2-subgroups of GL$(2, F)$ when $u(F) = 1$ $\{v = u(F(\iota))\}$ (see Table 9).

Let τ be a primitive cube root of 1 in $\mathbb{C}$ and consider the following subfields of $\mathbb{C}$: $\mathbb{Q}$, $\mathbb{Q}(\sqrt{-2})$, $\mathbb{Q}(\sqrt{-3}) = \mathbb{Q}(\tau)$. We obtain the following isomorphism classes of maximal 2-subgroups:

GL$(2, \mathbb{Q})$: $P_{2102} \approx D_8$;

GL$(2, \mathbb{Q}(\sqrt{-2}))$: $P_{3112} \approx SD_{16}$;

GL$(2, \mathbb{Q}(\sqrt{-3}))$: $P_{2102} \approx D_8$ and $P_{2122} \approx Q_8$.

The last is perhaps the simplest example of a violation of Sylow's Theorem.

Table 9

Group	OSA	Gens	2-overgroups
P_{k102} $(2 \le k < v-1)$	Z_2	y_{k+1}	$P_{k+1,1,0,2}$
$P_{v-1,1,0,2}$ $(3 \le v < \infty,\ \kappa_v \in F)$	Z_2	y_v	P_{v102}
$P_{v-1,1,0,2}$ $(3 \le v < \infty,\ \iota\kappa_v \in F)$	Z_2	ιy_v	P_{v112}
P_{v102} $(2 \le v < \infty,\ \kappa_v \in F)$	Z_2	$(1/\kappa_{v+1})y_{v+1}$ [1]	-
P_{v112} $(3 \le v < \infty,\ \iota\kappa_v \in F)$	(1)	-	-
P_{2122} $(v = 2,\ s(F) = 2)$	Σ_3	ω, $(1/\kappa_3)y_3$	-
P_{2122} $(v = 3,\ \kappa_3 \in F,\ s(F) = 2)$	Σ_3	ω, y_3	3 groups iso. to P_{3122} [2]
P_{2122} $(v = 3,\ \iota\kappa_3 \in F)$	Σ_3	ω, ιy_3	3 groups iso. to P_{3112} [2]
P_{2122} $(v > 3,\ s(F) = 2)$	Σ_3	ω, y_3	3 groups iso. to P_{3122} [2]
P_{k122} $(2 < k < v-1,\ s(F) = 2)$	Z_2	y_{k+1}	$P_{k+1,1,2,2}$
$P_{v-1,1,2,2}$ $(3 < v < \infty,\ \kappa_v \in F,\ s(F) = 2)$	Z_2	y_v	P_{v122}
$P_{v-1,1,2,2}$ $(3 < v < \infty,\ \iota\kappa_v \in F)$	Z_2	ιy_v	iso. P_{v112} [3]
P_{v122} $(3 \le v < \infty,\ \kappa_v \in F,\ s(F) = 2)$	Z_2	$(1/\kappa_{v+1})y_{v+1}$	-

[1] $(1/\kappa_{v+1})y_{v+1} = \begin{pmatrix} 1 & (-1)^{v+1}\sigma_v/(\kappa_v+1) \\ (-1)^v\sigma_v/(\kappa_v+1) & 1 \end{pmatrix}$ lies in $\mathrm{GL}(2, F)$, when $\kappa_v \in F$.

[2] The 3 overgroups come from the Sylow 2-subgroups of Σ_3 .

[3] The overgroup is precisely P_{v112} if we take $\xi = \iota(-1)^{u+1}\sigma_v$ and $\eta = \iota\kappa_v$ in x'' .

12. SL(2, F)

Every finite nonabelian 2-subgroup of $\mathrm{SL}(2, F)$ is isomorphic to some P_{k122} $(\approx Q_{2^{k+1}})$. The same analysis is used as in Section 4 and the notation follows that of Section 11. If $|F| < \infty$ and $u = 1$, then $\iota\kappa_v \in F$.

Table 10

Group	OSA	Gens	Overgroups	Conjugacy Set
$\underline{u = 1}$				
P_{2122} $(v = 2,\ s(F) = 2)$	Z_3	$(1/2\kappa_3)y_3\omega$	-	$F^\times/\langle F^{\times 2},\ 2\rangle$
P_{2122} $(v = 3,\ \kappa_3 \in F,\ s(F) = 2)$	Σ_3	$\kappa_3\omega,\ y_3$	3 groups iso.[1] to P_{3122}	$F^\times/F^{\times 2}$
P_{2122} $(v = 3,\ \iota\kappa_3 \in F)$	Z_3	$\kappa_3 y_3\omega$	-	$F^\times/\langle F^{\times 2},\ 2\rangle$
P_{2122} $(v < 3,\ s(F) = 2)$	Σ_3	$\kappa_3\omega,\ y_3$	3 groups iso.[1] to P_{3122}	$F^\times/F^{\times 2}$
P_{k122} $(2 < k < v\text{-}1,\ s(F) = 2)$	Z_2	y_{k+1}	$P_{k+1,1,2,2}$	$F^\times/F^{\times 2}$
$P_{v-1,1,2,2}$ $(3 < v < \infty,\ \kappa_v \in F,\ s(F) = 2)$	Z_2	y_v	P_{v122}	$F^\times/F^{\times 2}$
$P_{v-1,1,2,2}$ $(3 < v < \infty,\ \iota\kappa_v \in F)$	(1)	-	-	$F^\times/\langle F^{\times 2},\ -1\rangle$
P_{v122} $(3 \le v < \infty,\ \kappa_v \in F,\ s(F) = 2)$	(1)	-	-	$F^\times/\langle F^{\times 2},\ 1+\kappa_v\rangle$
$\underline{u = 2}$				
P_{2122}	Z_3	$(1/(1+\iota))\omega(\alpha_3 y_3)$	-	$F^\times/\langle F^{\times 2},\ \iota\rangle$
$\underline{u > 2}$				
P_{2122}	Σ_3	$\omega,\ y_3$	3 groups iso.[1] to P_{3122}	$F^\times/F^{\times 2}$
P_{k122} $(3 \le k < u)$	Z_2	y_{k+1}	$P_{k+1,1,2,2}$	$F^\times/F^{\times 2}$
P_{u122} $(u < \infty)$	(1)	-	-	$F^\times/\langle F^{\times 2},\ F_2\rangle$

[1] These 3 overgroups come from the Sylow 2-subgroups of Σ_3 .

13. PGL(2, F)

Finite nonabelian 2-subgroups of PGL(2, F) are isomorphic to P_{k102} . One works with preimages in GL . A preimage of a copy of P_{k102} is conjugate to a subgroup $Q_{k+1}(\gamma,\ \delta) \le$ GL , with $\gamma,\ \delta \in F$ and $\gamma^2 + \delta^2 \ne 0$. Here we define $x(\gamma,\ \delta) = \begin{pmatrix}\gamma & \delta \\ \delta & -\gamma\end{pmatrix}$. Then we have (p. 32)

Table 11

Group		OSA	Gens	Overgroups	Conjugacy Set
P_{k102}	$(2 \leq k < u-1)$	Z_2	y_{k+2}	$P_{k+1,1,0,2}$	$F^{\times}/F^{\times 2}$
$P_{u-1,1,0,2}$	$(3 \leq u < \infty)$	Z_2	$\alpha_{u+1}y_{u+1}$	P_{u102}	$F^{\times}/F^{\times 2}$
P_{u102}	$(2 \leq u < \infty)$	(1)	-	-	$F^{\times}/\langle F^{\times 2}, F_2\rangle$
P_{k102}	$(u = 1, 2 \leq k < v-2)$	Z_2	y_{k+2}	$P_{k+1,1,0,2}$	$(F^2+F^2)^{\times}/F^{\times 2}$
$P_{v-2,1,0,2}$	$(u = 1, 4 \leq v < \infty, \kappa_v \in F)$	Z_2	y_v	$P_{v-1,1,0,2}$	$(F^2+F^2)^{\times}/F^{\times 2}$
$P_{v-2,1,0,2}$	$(u = 1, 4 \leq v < \infty, \iota\kappa_v \in F)$	Z_2	ιy_v	$P_{v-1,1,0,2}$	$(F^2+F^2)^{\times}/F^{\times 2}$
$P_{v-1,1,0,2}$	$(u = 1, 3 \leq v < \infty, \kappa_v \in F)$	Z_2	$(1/\kappa_{v+1})y_{v+1}$	P_{v102}	$(F^2+F^2)^{\times}/F^{\times 2}$
$P_{v-1,1,0,2}$	$(u = 1, 3 \leq v < \infty, \iota\kappa_v \in F)$	(1)	-	-	$(F^2+F^2)^{\times}/\langle F^{\times 2}, -1\rangle$
P_{v102}	$(u = 1, 3 \leq v < \infty, \kappa_v \in F)$	(1)	-	-	$(F^2+F^2)^{\times}/\langle F^{\times 2}, 1+\kappa_v\rangle$
P_{2102}	$(u = 1, v = 2)$	(1)	-	-	$(F^2+F^2)^{\times}/\langle F^{\times 2}, 2\rangle$

$$Q_k(\gamma,\ \delta) = F^\times . \langle x(\gamma,\ \delta),\ y_1,\ \ldots,\ y_k\rangle \quad (3 \le k < u+1)\ ,$$

$$Q_{u+1}(\gamma,\ \delta) = F^\times . \left\langle x(\gamma,\ \delta),\ y_1,\ \ldots,\ y_u,\ y_{u+1}^{(1)} = \alpha_{u+1}y_{u+1}\right\rangle \quad (2 \le u < \infty)\ ,$$

$$Q_k(\gamma,\ \delta) = F^\times . \langle x(\gamma,\ \delta),\ y_1,\ \ldots,\ y_k\rangle \quad (u = 1,\ 3 \le k < v)\ ,$$

$$Q_v(\gamma,\ \delta) = F^\times . \langle x(\gamma,\ \delta),\ y_1,\ \ldots,\ y_v\rangle \quad (u = 1,\ 3 \le v < \infty,\ \kappa_v \in F)\ ,$$

$$Q_v(\gamma,\ \delta) = F^\times . \langle x(\gamma,\ \delta),\ y_1,\ \ldots,\ y_{v-1},\ \iota y_v\rangle \quad (u = 1,\ 3 \le v < \infty,\ \iota\kappa_v \in F)\ ,$$

and

$$Q_{v+1}(\gamma,\ \delta) = F^\times . \langle x(\gamma,\ \delta),\ y_1,\ \ldots,\ y_v,\ (1/\kappa_{v+1})y_{v+1}\rangle \quad (u = 1,\ 2 \le v < \infty,\ \kappa_v \in F)\ .$$

The $Q_k(\gamma,\ \delta)$ $(k > 2)$ have one abelian maximal subgroup which contains $F^\times I$ and conjugacy classes are determined by $-\det$ of elements of $Q_k(\gamma,\ \delta)$ which lie outside this maximal subgroup. Thus $-\det(x(\gamma,\ \delta)) = \gamma^2 + \delta^2$. Write $(F^2+F^2)^\times$ for the subgroup of $F^\times$ composed of elements which can be written as the sum of two squares. Note that if $u \ge 2$ or if $|F| < \infty$, then $(F^2+F^2)^\times = F^\times$. The conjugacy set of $P_{k102} \le \mathrm{PGL}$ is the set of possible values for $-\det(x(\gamma,\ \delta)) = \gamma^2 + \delta^2$, modulo determinants of elements in the maximal subgroup of Q_{k+1}. We write $\mathrm{OSA}(P_{k102}) = N_{\mathrm{PGL}}(P_{k102})/P_{k102} \approx N_{\mathrm{GL}}(Q_{k+1})/Q_{k+1}$, and we give the generators of this group in GL (see Table 11).

14. PSL(2, F)

We continue the notation of Section 13 and look to see which of the $Q_k(\gamma,\ \delta) \le F^\times . \mathrm{SL}$, or again we look at those $Q_k(\gamma,\ \delta)$'s all of whose elements have determinants which are squares in F. Thus we can confine our attention to the conjugacy class in PGL which contains $Q_k(1,\ 0)$. The conjugacy classes of the $Q_k(1,\ 0)$ in SL are then determined by the determinant of the conjugating element in GL and the conjugacy set is the quotient of $F^\times$ by the subgroup of determinants of elements in $N_{\mathrm{GL}}(Q_k(1,\ 0))$. We write

$$\mathrm{OSA}(P_{k102}) = N_{\mathrm{PSL}}(P_{k102})/P_{k102} \approx N_{F^\times.\mathrm{SL}}(Q_{k+1}(1,\ 0))/Q_{k+1}(1,\ 0)\ .$$

Table 12

Group		OSA	Gens	Overgroups	Conjugacy Set
P_{k102}	$(2 \leq k < u-1)$	Z_2	y_{k+2}	$P_{k+1,1,0,2}$	$F^{\times}/F^{\times 2}$
$P_{u-1,1,0,2}$	$(3 \leq u < \infty)$	(1)	-	-	$F^{\times}/\langle F^{\times 2}, F_2\rangle$
P_{k102}	$(u = 1,\ 2 \leq k < v-2)$	Z_2	y_{k+2}	$P_{k+1,1,0,2}$	$F^{\times}/F^{\times 2}$
$P_{v-1,1,0,2}$	$(u = 1,\ 4 \leq v < \infty,\ \kappa_v \in F)$	Z_2	y_v	$P_{v-1,1,0,2}$	$F^{\times}/F^{\times 2}$
$P_{v-2,1,0,2}$	$(u = 1,\ 4 \leq v < \infty,\ \iota\kappa_v \in F)$	(1)	-	-	$F^{\times}/\ F^{\times 2}, -1$
$P_{v-1,1,0,2}$	$(u = 1,\ 3 \leq v < \infty,\ \kappa_v \in F)$	(1)	-	-	$F^{\times}/\langle F^{\times 2}, 1+\kappa_v\rangle$

15. U(2, F)

Let $- : F \to F$, $\phi \mapsto \overline{\phi}$ be the involution of F , $N : F^{\times} \to N(F^{\times}) \leq F^{\times}$ be the norm homomorphism and let $\overline{F}$ be the subfield of F left invariant by $-$. A close look at the effect of $-$ on the group F_2 of 2 power roots of 1 in F is required. We distinguish the following cases:

(15.1) .1 : $1 < u$, $\overline{\alpha_k} = \alpha_k$ $(1 < k < u+1)$,

.2 : $1 < u$, $\overline{\alpha_k} = \alpha_k^{-1}$ $(1 < k < u+1)$,

.3 : $2 < u < \infty$, $\overline{\alpha_k} = \alpha_k^{-1}$ $(1 < k < u)$, $\overline{\alpha_u} = -\alpha_u^{-1}$,

.4 : $2 < u < \infty$, $\overline{\alpha_k} = \alpha_k$ $(1 < k < u)$, $\overline{\alpha_u} = -\alpha_u$,

.5 : $u = 1$.

If $u < \infty$ in case .2, then characteristic of F is 0 .

If $|F| < \infty$, we can suppose $F = \mathrm{F}_{q^2}$. If q has form $4n + 3$, we have case .3. If q has form $4n + 1$, we have case .4.

LEMMA 15.2. *U(2, F) contains the quaternion group Q_8 $(\approx P_{2122})$ iff -1 is the sum of squares of two antisymmetric elements of F ; in this case we write $\overline{s}(F) = 2$. If $-1 = \xi^2 + \eta^2$ with $\overline{\xi} = -\xi$ and $\overline{\eta} = -\eta$, we write $x'' = \begin{pmatrix} \xi & \eta \\ \eta & -\xi \end{pmatrix}$ and then $P_{2122} = \langle y_2, x'' \rangle$. The similarity automorphism*

$\omega : y_2 \to x''$, $x'' \to y_2$ *of* P_{2122} *is realised by the matrix* $\begin{pmatrix} \xi & \eta+1 \\ \eta-1 & -\xi \end{pmatrix}$ *of determinant* 2 . *If* $u \geq 2$ *and* $\bar{\iota} = -\iota$, *then* $-1 = \iota^2 + 0^2$. *If* $u \geq 2$ *and* $\bar{\iota} = \iota$, *then* $-1 = (\zeta-(1/4\zeta))^2 + (\iota\zeta+(\iota/4\zeta))^2$, *where* ζ *is any nonzero antisymmetric element of* F .

(15.3) It is convenient to know which isomorphic types of nonabelian 2-subgroups exist in each case. Here $u = u(F)$, $v = u(F(\iota))$ and $w = u(\overline{F}(\iota))$.

.1 : We have P_{k102} and P_{k122} $(2 \leq k < u+1)$.

.2 : If $u = \infty$, we have all P_{klm2} .

If $u < \infty$, we have all subgroups of $P_{u+1,u,1,2}$.

.3 : We have all subgroups of $P_{u,u-1,1,2}$.

.4 : We have all subgroups of P_{u112} .

.5 : We have P_{k102} $(2 \leq k < w)$ and P_{k122} $(2 \leq k < w, \bar{s}(F) = 2)$.

Also if $2 \leq w < \infty$ and $\kappa_w \in F$, we have P_{w102} and P_{w122} $(\bar{s}(F) = 2)$.

(15.4) From this, it is becoming clear that many subcases will need to be considered. A full list of such subcases is supplied in Table 13.

(15.5) To obtain conjugacy classes we must consider the following subgroup of GL :

$$V = \{g \in \mathrm{GL} \mid g\bar{g}^t = \lambda I\} .$$

The element $\lambda \in \overline{F}$. U is the normal subgroup of V consisting of those elements g with $\lambda = 1$. Also $\mathrm{SU} \triangleleft V$.

LEMMA 15.6. *Let* G *be an absolutely irreducible group of matrices contained in* U .

(i) If $g \in \mathrm{GL}$, *then* $G^g \leq \mathrm{U}$ *iff* $g \in V$.

(ii) The conjugacy classes of G *in* U *are in one-to-one correspondence with the cosets of* $N_V(G)$. U *in* V .

(iii) If $G \leq \mathrm{SU}$, *then the conjugacy classes of* G *in* SU *are in one-to-one correspondence with the cosets of* $N_V(G)$. SU *in* V .

Table 13

.1 : $1 < u$, $\bar{\alpha}_k = \alpha_k$ $(1 < k < u+1)$

.11 : $u = 2$

.111 : $2 \in N(F) \Leftrightarrow \iota \in N(F)$.

.112 : $2 \notin N(F) \Leftrightarrow \iota \notin N(F)$.

.12 : $2 < u < \infty$.

.121 : $\alpha_u \in N(F)$.

.122 : $\alpha_u \notin N(F)$.

.13 : $u = \infty$.

.2 : $1 < u$, $\bar{\alpha}_k = \alpha_k^{-1}$ $(1 < k < u+1)$.

.21 : $u = 2$.

.22 : $2 < u < \infty$.

.23 : $u = \infty$.

.3 : $2 < u < \infty$, $\bar{\alpha}_k = \alpha_k^{-1}$ $(1 < k < u)$, $\bar{\alpha}_u = -\alpha_u^{-1}$.

.31 : $u = 3$.

.32 : $3 < u < \infty$.

.4 : $2 < u < \infty$, $\bar{\alpha}_k = \alpha_k$ $(1 < k < u)$, $\bar{\alpha}_u = -\alpha_u$.

.41 : $u = 3$.

.42 : $3 < u < \infty$.

.5 : $u = 1$.

.51 : $w = 2$.

.511 : $2 \in N(F)$.

.512 : $2 \notin N(F)$.

.52 : $w = 3$.

.521 : $\kappa_3 \in \bar{F}$.

.5211 : $\kappa_3 + 1 \in N(F)$.

.5212 : $\kappa_3 + 1 \notin N(F)$.

.522 : $\iota\kappa_3 \in \bar{F}$.

.5221 : $-1 \in N(F)$.

.5222 : $-1 \notin N(F)$.

.53 : $3 < w < \infty$.

.531 : $\kappa_w \in \bar{F}$.

.5311 : $\kappa_w + 1 \in N(F)$.

.5312 : $\kappa_w + 1 \notin N(F)$.

.532 : $\iota\kappa_w \in \bar{F}$.

.5321 : $-1 \in N(F)$.

.5322 : $-1 \notin N(F)$.

.54 : $w = \infty$.

(15.7) We have the following picture of subgroups of V :

$$V = \left\{ \begin{pmatrix} \lambda\bar{\delta} & -\lambda\bar{\varepsilon} \\ \varepsilon & \delta \end{pmatrix} \mid \delta\bar{\delta}+\varepsilon\bar{\varepsilon} \neq 0,\ \lambda\bar{\lambda} = 1 \right\} ;$$

$$U = \left\{ \begin{pmatrix} \lambda\bar{\delta} & -\lambda\bar{\varepsilon} \\ \varepsilon & \delta \end{pmatrix} \mid \delta\bar{\delta}+\varepsilon\bar{\varepsilon} = 1,\ \lambda\bar{\lambda} = 1 \right\} \trianglelefteq V ;$$

$$A = \left\{ \begin{pmatrix} \bar{\delta} & -\bar{\varepsilon} \\ \varepsilon & \delta \end{pmatrix} \mid \delta\bar{\delta}+\varepsilon\bar{\varepsilon} \neq 0 \right\} \trianglelefteq V ;$$

$$SU = A \cap U = \left\{ \begin{pmatrix} \bar{\delta} & -\bar{\varepsilon} \\ \varepsilon & \delta \end{pmatrix} \mid \delta\bar{\delta}+\varepsilon\bar{\varepsilon} = 1 \right\} \trianglelefteq V ;$$

$$B = \left\{ \begin{pmatrix} \lambda & 0 \\ 0 & 1 \end{pmatrix} \mid \lambda\bar{\lambda} = 1 \right\} \leq V .$$

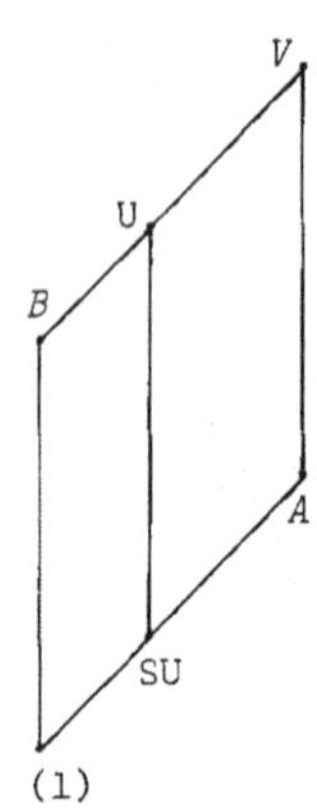

We have the following group epimorphism:

$$\alpha : V \to \left(N(F)+N(F)\right)^{\times} , \quad \begin{pmatrix} \lambda\bar{\delta} & -\lambda\bar{\varepsilon} \\ \varepsilon & \delta \end{pmatrix} \mapsto \delta\bar{\delta} + \varepsilon\bar{\varepsilon} .$$

Here $\left(N(F)+N(F)\right)^{\times}$ is the subgroup of $F^{\times}$ of all elements of the form $\delta\bar{\delta} + \varepsilon\bar{\varepsilon}$. α leads to the isomorphism:

$$\xi : V/F^{\times} . U \approx \left(N(F)+N(F)\right)^{\times}/N(F)^{\times} .$$

We also have the group epimorphism:

$$\beta : V \to N^{-1}(1) , \quad \begin{pmatrix} \lambda\bar{\delta} & -\lambda\bar{\varepsilon} \\ \varepsilon & \delta \end{pmatrix} \mapsto \lambda .$$

If $\lambda \in N^{-1}(1)$ (that is, $\lambda\bar{\lambda} = 1$) , we can write $\lambda = \bar{\mu}/\mu$: for example, take $\mu = \bar{\lambda} + 1$ (or $\mu = \zeta$, if $\lambda = -1$ and $\bar{\zeta} = -\zeta \neq 0$) and then μ is well determined up to a factor from $\bar{F}^{\times}$. (From this it readily follows that $V = F^{\times} . A$.) This gives the epimorphism $\gamma : N^{-1}(1) \to N(F)^{\times}/\bar{F}^{\times 2}$, $\lambda \mapsto \mu\bar{\mu}\bar{F}^{\times 2}$ (γ induces the isomorphism $\gamma_1 : N^{-1}(1)/N^{-1}(1)^2 \to N(F)^{\times}/\bar{F}^{\times 2}$). We get two homomorphisms α_1 and $\beta_1 : V \to \left(N(F)+N(F)\right)^{\times}/\bar{F}^{\times 2}$, the former derived from α and the latter from the composite of β and γ . The product (or sum) of α_1 and β_1 (given by multiplication of images, elementwise) gives the isomorphism:

$$\eta : V/F^{\times} . SU \to \left(N(F)+N(F)\right)^{\times}/\bar{F}^{\times 2} .$$

As $N_V(G) . U$ and $N_V(G) . SU$ (if $G \leq SU$) are not much larger than $F^{\times} . U$

and $F^\times$. SU respectively, it is now a simple matter to describe the conjugacy sets for the different G ($\le$ U) which arise. We define $\mathrm{OSA}(G) = N_{\mathrm{U}}(G)/G$ and give its generators in terms of matrices in U . We also give the generators of $N_V(G)$. U/$F^\times$. U = L in terms of matrices.

We always have that

$$\left(\overline{F}^2+\overline{F}^2\right)^\times \le \left(N(F)+N(F)\right)^\times \le \overline{F}^\times .$$

Hence if $\overline{F} = \overline{F}^2 + \overline{F}^2$ we may replace $\left(N(F)+N(F)\right)^\times$ by $\overline{F}^\times$. This condition is true if $u(\overline{F}) \ge 2$ or if $|\overline{F}| < \infty$. Thus it holds in particular in cases .1 and .4. Also, if $|F| < \infty$, then $N(F)^\times = \overline{F}^\times$.

Table 14

Case	Group	OSA	OSA Gens	2-overgroups	L Gens	Conjugacy Set
.111	P_{2102}	Z_2	$\beta\alpha_3 y_3$ [1]	-	-	$\frac{\overline{F}^\times}{N(F)^\times}$
.111	P_{2122}	Σ_3	$\beta\alpha_3 y_3$, [1] $(\iota/\beta(1+\iota))\omega$	-	-	$\frac{\overline{F}^\times}{N(F)^\times}$
.112	P_{2102}	(1)	-	-	$\alpha_3 y_3$	$\frac{\overline{F}^\times}{\langle N(F)^\times,\iota\rangle}$
.112	P_{2122}	Z_3	$(\alpha_3/(1+\iota))y_3\omega$	-	$\alpha_3 y_3$	$\frac{\overline{F}^\times}{\langle N(F)^\times,\iota\rangle}$
.12 & .13	P_{k102} $(2 \le k < u)$	Z_2	y_{k+1}	$P_{k+1,1,0,2}$	-	$\frac{\overline{F}^\times}{N(F)^\times}$
.121	P_{u102}	Z_2	$\nu\alpha_{u+1}y_{u+1}$ [7]	-	-	$\frac{\overline{F}^\times}{N(F)^\times}$
.122	P_{u102}	(1)	-	-	$\alpha_{u+1}y_{u+1}$	$\frac{\overline{F}^\times}{\langle N(F)^\times,\alpha_u\rangle}$
.12 & .13	P_{2122}	Σ_3	y_3, $\kappa_3\omega$	3 groups iso.[2] to P_{3122}	-	$\frac{\overline{F}^\times}{N(F)^\times}$
.12 & .13	P_{k122} $(3 \le k < u)$	Z_2	y_{k+1}	$P_{k+1,1,2,2}$	-	$\frac{\overline{F}^\times}{N(F)^\times}$
.121	P_{u122}	Z_2	$\nu\alpha_{u+1}y_{u+1}$	-	-	$\frac{\overline{F}^\times}{N(F)^\times}$

Table 14 (continued)

Case	Group	OSA	OSA Gens	2-overgroups	L Gens	Conjugacy Set
.122	P_{u122}	(1)	-	-	$\alpha_{u+1}y_{u+1}$	$\overline{F}^{\times} / \langle N(F)^{\times}, \alpha_u \rangle$
.21	P_{2102}	Z_2	$\alpha_3 y_3$	P_{2202}	-	$(N(F)+N(F))^{\times} / N(F)^{\times}$
.21	P_{2202}	Σ_3	ω, $\alpha_3 y_3$	3 groups iso.[2] to P_{3212}	-	$(N(F)+N(F))^{\times} / N(F)^{\times}$
.21	P_{3212}	(1)	-	-	-	$(N(F)+N(F))^{\times} / N(F)^{\times}$
.21	P_{2122}	Σ_3	$\alpha_3 y_3$, $(1/1+\iota)\omega$	P_{2202}	-	$(N(F)+N(F))^{\times} / N(F)^{\times}$
.21.	P_{2222}	Z_2	$\alpha_3 y_3$	P_{3212}	-	$(N(F)+N(F))^{\times} / N(F)^{\times}$
.22 & .23	P_{2102}	Z_2	y_3	P_{2202}, P_{3102}, P_{3112}	-	$(N(F)+N(F))^{\times} / N(F)^{\times}$
.22 & .23	P_{2l02} $(2 \le l < u)$	Σ_3	y_3, ω	$P_{2,l+1,0,2}$, 3 groups iso.[2] to P_{3lm2} for each $m \in (0, 1)$	-	$(N(F)+N(F))^{\times} / N(F)^{\times}$
.22	P_{2u02}	Σ_3	y_3, ω	3 groups iso.[2] to P_{3u02}	-	$(N(F)+N(F))^{\times} / N(F)^{\times}$
.22 & .23	P_{kl02} $(3 \le k < u,$ $1 \le l < u)$	Z_2	y_{k+1}	$P_{k,l+1,0,2}$, $P_{k+1,l,m,2}$ for $m \in (0, 1)$	-	$(N(F)+N(F))^{\times} / N(F)^{\times}$
.22	P_{ku02} $(3 \le k < u)$	Z_2	y_{k+1}	$P_{k+1,u,0,2}$	-	$(N(F)+N(F))^{\times} / N(F)^{\times}$
.22	P_{ul02} $(1 \le l < u)$	Z_2	$\alpha_{u+1}y_{u+1}$	$P_{u,l+1,0,2}$	-	$(N(F)+N(F))^{\times} / N(F)^{\times}$
.22	P_{uu02}	Z_2	$\alpha_{u+1}y_{u+1}$	$P_{u+1,u,1,2}$	-	$(N(F)+N(F))^{\times} / N(F)^{\times}$

Table 14 (continued)

Case	Group	OSA	OSA Gens	2-overgroups	L Gens	Conjugacy Set
.22 & .23	P_{kl12} ($3 \le k < u+1$, $1 \le l < u$)	(1)	-	$P_{k,l+1,0,2}$	-	$\frac{(N(F)+N(F))^{\times}}{N(F)^{\times}}$
.22	$P_{u+1,u,1,2}$	(1)	-	-	-	$\frac{(N(F)+N(F))^{\times}}{N(F)^{\times}}$
.22 & .23	P_{2122}	Σ_3	y_3, $(1/1+\iota)\omega$	P_{2202}, 3 groups iso.[2] to P_{31m2} for each $m \in (1, 2)$	-	$\frac{(N(F)+N(F))^{\times}}{N(F)^{\times}}$
.22 & .23	P_{kl22} ($\neq P_{2122}$, $2 \le k < u$, $1 \le l < u$)	Z_2	y_{k+1}	$P_{k,l+1,0,2}$, $P_{k+1,l,2,2}$, 1 group iso.[3] to $P_{k+1,l,1,2}$	-	$\frac{(N(F)+N(F))^{\times}}{N(F)^{\times}}$
.22	P_{ku22} ($2 \le k < u$)	Z_2	y_{k+1}	$P_{k+1,u,2,2}$	-	$\frac{(N(F)+N(F))^{\times}}{N(F)^{\times}}$
.22	P_{ul22} ($1 \le l < u$)	Z_2	$\alpha_{u+1}y_{u+1}$	$P_{u,l+1,0,2}$	-	$\frac{(N(F)+N(F))^{\times}}{N(F)^{\times}}$
.22	P_{uu22}	Z_2	$\alpha_{u+1}y_{u+1}$	$P_{u+1,u,1,2}$	-	$\frac{(N(F)+N(F))^{\times}}{N(F)^{\times}}$
.31	P_{2102}	Z_2	$\alpha_3 y_3$	P_{2202}	-	$\frac{(N(F)+N(F))^{\times}}{N(F)^{\times}}$
.31	P_{2202}	Σ_3	ω, $\alpha_3 y_3$	3 groups iso.[2] to P_{3212}	-	$\frac{(N(F)+N(F))^{\times}}{N(F)^{\times}}$
.31	P_{2122}	Σ_3	$(1/1+\iota)\omega$, $\alpha_3 y_3$	P_{2202}	-	$\frac{(N(F)+N(F))^{\times}}{N(F)^{\times}}$
.31	P_{2222}	Z_2	$\alpha_3 y_3$	P_{3212}	-	$\frac{(N(F)+N(F))^{\times}}{N(F)^{\times}}$
.31 & .32	$P_{u,u-1,1,2}$	(1)	-	-	-	$\frac{(N(F)+N(F))^{\times}}{N(F)^{\times}}$
.32	P_{2102}	Z_2	y_3	P_{2202}, P_{3102}, P_{3112}	-	$\frac{(N(F)+N(F))^{\times}}{N(F)^{\times}}$

Table 14 (continued)

Case	Group	OSA	OSA Gens	2-overgroups	L Gens	Conjugacy Set
.32	P_{2l02} $(2 \le l < u-1)$	Σ_3	$\omega,\ y_3$	$P_{2,l+1,0,2}$, 3 groups iso.[2] to P_{3lm2} for each $m \in (0,\ 1)$	-	$\frac{(N(F)+N(F))^{\times}}{N(F)^{\times}}$
.32	$P_{2,u-1,0,2}$	Σ_3	$\omega,\ y_3$	3 groups iso.[2] to $P_{3,u-1,0,2}$	-	$\frac{(N(F)+N(F))^{\times}}{N(F)^{\times}}$
.32	P_{kl02} $(3 \le k < u-1,$ $1 \le l < u-1)$	Z_2	y_{k+1}	$P_{k,l+1,0,2}$, $P_{k+1,l,m,2}$ for each $m \in (0,\ 1)$	-	$\frac{(N(F)+N(F))^{\times}}{N(F)^{\times}}$
.32	$P_{u-1,l,0,2}$ $(1 \le l < u-1)$	Z_2	$\alpha_u y_u$	$P_{u-1,l+1,0,2}$	-	$\frac{(N(F)+N(F))^{\times}}{N(F)^{\times}}$
.32	$P_{k,u-1,0,2}$ $(3 \le k < u-1)$	Z_2	y_{k+1}	$P_{k+1,u-1,0,2}$	-	$\frac{(N(F)+N(F))^{\times}}{N(F)^{\times}}$
.32	$P_{u-1,u-1,0,2}$	Z_2	$\alpha_u y_u$	$P_{u,u-1,1,2}$	-	$\frac{(N(F)+N(F))^{\times}}{N(F)^{\times}}$
.32	P_{kl12} $(3 \le k < u,$ $1 \le l < u-1)$	(1)	-	$P_{k,l+1,0,2}$	-	$\frac{(N(F)+N(F))^{\times}}{N(F)^{\times}}$
.32	P_{2122}	Σ_3	$(1/1+\iota)\omega,\ y_3$	P_{2202}, 3 groups iso.[2] to P_{31m2} for each $m \in (1,\ 2)$	-	$\frac{(N(F)+N(F))^{\times}}{N(F)^{\times}}$
.32	P_{kl22} $(\ne P_{2122},$ $2 \le k < u-1,$ $1 \le l < u-1)$	Z_2	y_{k+1}	$P_{k,l+1,0,2}$, $P_{k+1,l,2,2}$ and 1 group iso.[3] to $P_{k+1,l,1,2}$	-	$\frac{(N(F)+N(F))^{\times}}{N(F)^{\times}}$
.32	$P_{k,u-1,2,2}$ $(2 \le k < u-1)$	Z_2	y_{k+1}	$P_{k+1,u-1,2,2}$	-	$\frac{(N(F)+N(F))^{\times}}{N(F)^{\times}}$
.32	$P_{u-1,l,2,2}$ $(1 \le l < u-1)$	Z_2	$\alpha_u y_u$	$P_{u-1,l+1,0,2}$	-	$\frac{(N(F)+N(F))^{\times}}{N(F)^{\times}}$

Table 14 (continued)

Case	Group	OSA	OSA Gens	2-overgroups	L Gens	Conjugacy Set
.32	$P_{u-1,u-1,2,2}$	Z_2	$\alpha_u y_u$	$P_{u,u-1,1,2}$	-	$\frac{(N(F)+N(F))^{\times}}{N(F)^{\times}}$
.41	P_{2102}	Z_2	ιy_3	P_{3112}	-	$\frac{\overline{F}^{\times}}{N(F)^{\times}}$
.41	P_{2122}	Σ_3	$\iota\kappa_3\omega$, ιy_3	3 groups iso.[2] to P_{3112}	-	$\frac{\overline{F}^{\times}}{N(F)^{\times}}$
.41 & .42	P_{u112}	(1)	-	-	-	$\frac{\overline{F}^{\times}}{N(F)^{\times}}$
.42	P_{2102}	Z_2	y_3	P_{3102}	-	$\frac{\overline{F}^{\times}}{N(F)^{\times}}$
.42	P_{k102} $(3 \le k < u-1)$	Z_2	y_{k+1}	$P_{k+1,1,0,2}$	-	$\frac{\overline{F}^{\times}}{N(F)^{\times}}$
.42	$P_{u-1,1,0,2}$	Z_2	ιy_u	P_{u112}	-	$\frac{\overline{F}^{\times}}{N(F)^{\times}}$
.42	P_{2122}	Σ_3	$\kappa_3\omega$, y_3	3 groups iso.[2] to P_{3122}	-	$\frac{\overline{F}^{\times}}{N(F)^{\times}}$
.42	P_{k122} $(3 \le k < u-1)$	Z_2	y_{k+1}	$P_{k+1,1,2,2}$	-	$\frac{\overline{F}^{\times}}{N(F)^{\times}}$
.42	$P_{u-1,1,2,2}$	Z_2	ιy_u	P_{u112}	-	$\frac{\overline{F}^{\times}}{N(F)^{\times}}$
.53 & .54	P_{k102} $(2 \le k < w-1)$	Z_2	y_{k+1}	$P_{k+1,1,0,2}$	-	$\frac{(N(F)+N(F))^{\times}}{N(F)^{\times}}$
.521 & .531	$P_{w-1,1,0,2}$	Z_2	y_w	P_{w102}	-	$\frac{(N(F)+N(F))^{\times}}{N(F)^{\times}}$
.5221 & .5321	$P_{w-1,1,0,2}$	Z_2	$\varepsilon\iota y_w$ [6]	-	-	$\frac{(N(F)+N(F))^{\times}}{N(F)^{\times}}$
.5222 & .5322	$P_{w-1,1,0,2}$	(1)	-	-	ιy_w	$\frac{(N(F)+N(F))^{\times}}{\langle N(F)^{\times}, -1\rangle}$
.511	P_{2102}	Z_2	$(\delta/\kappa_3)y_3$ [4]	-	-	$\frac{(N(F)+N(F))^{\times}}{N(F)^{\times}}$

Table 14 (continued)

Case	Group	OSA	OSA Gens	2-overgroups	L gens	Conjugacy Set
.512	P_{2102}	(1)	-	-	$(1/\kappa_3)y_3$	$\frac{(N(F)+N(F))^\times}{\langle N(F)^\times,2\rangle}$
.5211 & .5311	P_{w102}	Z_2	$(\gamma\kappa_3/\kappa_{w+1})y_{w+1}$ [5]	-	-	$\frac{(N(F)+N(F))^\times}{N(F)^\times}$
.5212 & .5312	P_{w102}	(1)	-	-	$(1/\kappa_{w+1})y_{w+1}$	$\frac{(N(F)+N(F))^\times}{\langle N(F)^\times,\kappa_w+1\rangle}$
.511	P_{2122} $(\bar{s}(F)=2)$	Σ_3	$(\delta/\kappa_3)y_3$, $\delta\omega$ [4]	-	-	$\frac{(N(F)+N(F))^\times}{N(F)^\times}$
.512	P_{2122} $(\bar{s}(F)=2)$	Z_3	$(1/2\kappa_3)y_3\omega$	-	ω	$\frac{(N(F)+N(F))^\times}{\langle N(F)^\times,2\rangle}$
.521, .53 & .54	P_{2122} $(\bar{s}(F)=2)$	Σ_3	$\kappa_3\omega$, y_3	3 groups iso.[2] to P_{3122}	-	$\frac{(N(F)+N(F))^\times}{N(F)^\times}$
.5221	P_{2122} $(\bar{s}(F)=2)$	Σ_3	$\varepsilon\iota y_3$, $\varepsilon\iota\kappa_3\omega$ [6]	-	-	$\frac{(N(F)+N(F))^\times}{N(F)^\times}$
.5222	P_{2122} $(\bar{s}(F)=2)$	Z_3	$\kappa_3 y_3\omega$	-	ω	$\frac{(N(F)+N(F))^\times}{\langle N(F)^\times,2\rangle}$
.53 & .54	P_{k122} $(2<k<w-1,$ $\bar{s}(F)=2)$	Z_2	y_{k+1}	$P_{k+1,1,2,2}$	-	$\frac{(N(F)+N(F))^\times}{N(F)^\times}$
.531	$P_{w-1,1,2,2}$ $(\bar{s}(F)=2)$	Z_2	y_w	P_{w122}	-	$\frac{(N(F)+N(F))^\times}{N(F)^\times}$
.5321	$P_{w-1,1,2,2}$ $(\bar{s}(F)=2)$	Z_2	$\varepsilon\iota y_w$ [6]	-	-	$\frac{(N(F)+N(F))^\times}{N(F)^\times}$
.5322	$P_{w-1,1,2,2}$ $(\bar{s}(F)=2)$	(1)	-	-	ιy_w	$\frac{(N(F)+N(F))^\times}{\langle N(F)^\times,-1\rangle}$
.5211 & .5311	P_{w122} $(\bar{s}(F)=2)$	Z_2	$(\gamma\kappa_3/\kappa_{w+1})y_{w+1}$ [5]	-	-	$\frac{(N(F)+N(F))^\times}{N(F)^\times}$
.5212 & .5312	P_{w122} $(\bar{s}(F)=2)$	(1)	-	-	$(1/\kappa_{w+1})y_{w+1}$	$\frac{(N(F)+N(F))^\times}{\langle N(F)^\times,\kappa_w+1\rangle}$

Table 14 (continued)

[1] $N(\beta) = -\iota = 1/\iota$.

[2] The three subgroups correspond to the Sylow 2-subgroups of Σ_3 .

[3] More precisely this group is $(P_{k+1,l,1,2})^{y_{k+2}}$ when $k < u - 1$ and $(P_{ul12})^{y_{u+1}^{(1)}}$ when $k = u - 1 < \infty$.

[4] $N(\delta) = \frac{1}{2}$.

[5] $N(\gamma) = \kappa_w + 1$.

[6] $N(\varepsilon) = -1$.

[7] $N(\nu) = 1/\alpha_u$.

16. SU(2, F)

As $SU = SL \cap U$, we can deduce which groups occur in SU from Sections 12 and 15. In Section 15 we saw that the conjugacy set of $G \leq SU$ is the set of cosets of $N_V(G) . SU$ in V and we saw that $V/F^\times . SU$ is isomorphic to $(N(F)+N(F))^\times/\overline{F}^{\times 2}$. Here we set $M = N_V(G) . SU/F^\times . SU$ and give generators of M in terms of matrices. We set $OSA(G) = N_{SU}(G)/G$.

Table 15

Case	Group	OSA	OSA Gens	2-overgroups	M Gens	Conjugacy Set
.11, .21 & .31	P_{2122}	Z_3	$(\alpha_3/1+\iota)y_3\omega$	-	$\alpha_3 y_3$	$\dfrac{(N(F)+N(F))^\times}{\langle \overline{F}^{\times 2}, 2\rangle}$
.12 & .13	P_{2122}	Σ_3	y_3, $\kappa_3\omega$	3 groups iso.[1] to P_{3122}	-	$\dfrac{\overline{F}^\times}{\overline{F}^{\times 2}}$
.12, .13, .22 & .23	P_{k122} $(3 \leq k < u)$	Z_2	y_{k+1}	$P_{k+1,1,2,2}$	-	$\dfrac{(N(F)+N(F))^\times}{\overline{F}^{\times 2}}$
.12	P_{u122}	(1)	-	-	$\alpha_{u+1}y_{u+1}$	$\dfrac{\overline{F}^\times}{\langle \overline{F}^{\times 2}, \alpha_u\rangle}$
.22, .23 & .32	P_{2122}	Σ_3	y_3, $(\alpha_3/1+\iota)\omega$	3 groups iso.[1] to P_{3122}	-	$\dfrac{(N(F)+N(F))^\times}{\overline{F}^{\times 2}}$

Table 15 (continued)

Case	Group	OSA	OSA Gens	2-overgroups	M Gens	Conjugacy Set
.32 & .42	P_{k122} $(3 \le k < u-1)$	Z_2	y_{k+1}	$P_{k+1,1,2,2}$	-	$\frac{(N(F)+N(F))^{\times}}{\overline{F}^{\times 2}}$
.32	$P_{u-1,1,2,2}$	(1)	-	-	y_u	$\frac{(N(F)+N(F))^{\times}}{\langle \overline{F}^{\times 2}, -1\rangle}$
.41	P_{2122}	Z_3	$\kappa_3 \omega y_3$	-	y_3	$\frac{\overline{F}^{\times}}{\langle \overline{F}^{\times 2}, 2\rangle}$
.42	P_{2122}	Σ_3	$\kappa_3\omega$, y_3	3 groups iso.[1] to P_{3122}	-	$\frac{\overline{F}^{\times}}{\overline{F}^{\times 2}}$
.42	$P_{u-1,1,2,2}$	(1)	-	-	y_u	$\frac{\overline{F}^{\times}}{\langle \overline{F}^{\times 2}, \alpha_{u-1}\rangle}$
.51	P_{2122} $(\overline{s}(F) = 2)$	Z_3	$(1/2\kappa_3)y_3\omega$	-	ω	$\frac{(N(F)+N(F))^{\times}}{\langle \overline{F}^{\times 2}, 2\rangle}$
.521 & .53	P_{2122} $(\overline{s}(F) = 2)$	Σ_3	$\kappa_3\omega$, y_3	3 groups iso.[1] to P_{3122}	-	$\frac{(N(F)+N(F))^{\times}}{\overline{F}^{\times 2}}$
.522	P_{2122} $(\overline{s}(F) = 2)$	Z_3	$\kappa_3\omega y_3$	-	ω	$\frac{(N(F)+N(F))^{\times}}{\langle \overline{F}^{\times 2}, 2\rangle}$
.53	P_{k122} $(3 \le k < w-1,$ $\overline{s}(F) = 2)$	Z_2	y_{k+1}	$P_{k+1,1,2,2}$	-	$\frac{(N(F)+N(F))^{\times}}{\overline{F}^{\times 2}}$
.531	$P_{w-1,1,2,2}$ $(\overline{s}(F) = 2)$	Z_2	y_w	P_{w122}	-	$\frac{(N(F)+N(F))^{\times}}{\overline{F}^{\times 2}}$
.532	$P_{w-1,1,2,2}$ $(\overline{s}(F) = 2)$	(1)	-	-	ιy_w	$\frac{(N(F)+N(F))^{\times}}{\langle \overline{F}^{\times 2}, -1\rangle}$
.521 & .531	P_{w122} $(\overline{s}(F) = 2)$	(1)	-	-	$(1/\kappa_{w+1})y_{w+1}$	$\frac{(N(F)+N(F))^{\times}}{\langle \overline{F}^{\times 2}, \kappa_w+1\rangle}$

[1] These 3 groups correspond to the Sylow 2-subgroups of Σ_3 .

17. PU(2, F)

We look to see which of the groups $Q_k(\gamma, \delta)$ $(k \ge 3)$, defined in Section 13

(p. 48)

Table 16

Case	Group	OSA	OSA gens	2-overgroups	sim. set	N gens	Conjugacy of $Q_{k+1}(\gamma, \delta)$ set
.1 & .2	P_{k102} $(2 \leq k < u-1)$	Z_2	y_{k+2}	$P_{k+1,1,0,2}$	$\frac{N(F)^{\times}}{\overline{F}^{\times 2}}$	-	$\frac{(N(F)+N(F))^{\times}}{N(F)^{\times}}$
.121	$P_{u-1,1,0,2}$	Z_2	$\beta\alpha_{u+1}y_{u+1}^{1}$	P_{u102}	$\frac{N(F)^{\times}}{\overline{F}^{\times 2}}$	-	$\frac{\overline{F}^{\times}}{N(F)^{\times}}$
.122	$P_{u-1,1,0,2}$	(1)	-	-	$\frac{N(F)^{\times}}{\overline{F}^{\times 2}}$	$\alpha_{u+1}y_{u+1}$	$\frac{\overline{F}^{\times}}{\langle N(F)^{\times}, \alpha_u\rangle}$
.111&& .121	P_{u102}	(1)	-	-	$\frac{N(F)^{\times}}{\langle \overline{F}^{\times 2}, \alpha_u\rangle}$	-	$\frac{\overline{F}^{\times}}{N(F)^{\times}}$
.22	$P_{u-1,1,0,2}$	Z_2	$\alpha_{u+1}y_{u+1}$	P_{u102}	$\frac{N(F)^{\times}}{\overline{F}^{\times 2}}$	-	$\frac{(N(F)+N(F))^{\times}}{N(F)^{\times}}$
.21	P_{2102}	(1)	-	-	$\frac{N(F)^{\times}}{\langle \overline{F}^{\times 2}, 2\rangle}$	-	$\frac{(N(F)+N(F))^{\times}}{N(F)^{\times}}$
.22	P_{u102}	(1)	-	-	$\frac{N(F)^{\times}}{\langle \overline{F}^{\times 2} \kappa_u + 1\rangle}$	-	$\frac{(N(F)+N(F))^{\times}}{N(F)^{\times}}$
.3 & .4	P_{k102} $(2 \leq k < u-2)$	Z_2	y_{k+2}	$P_{k+1,1,0,2}$	$\frac{N(F)^{\times}}{\overline{F}^{\times 2}}$	-	$\frac{(N(F)+N(F))^{\times}}{N(F)^{\times}}$
.32	$P_{u-2,1,0,2}$	Z_2	$\alpha_u y_u$	$P_{u-1,1,0,2}$	$\frac{N(F)^{\times}}{\overline{F}^{\times 2}}$	-	$\frac{(N(F)+N(F))^{\times}}{N(F)^{\times}}$
.31	P_{2102}	(1)	-	-	$\frac{N(F)^{\times}}{\langle \overline{F}^{\times 2}, 2\rangle}$	-	$\frac{(N(F)+N(F))^{\times}}{N(F)^{\times}}$

Table 16 (continued)

Case	Group	OSA	OSA gens	2-overgroups	sim. set	N gens	Conjugacy of $Q_{k+1}(\gamma,\delta)$ set
.32	$P_{u-1,1,0,2}$	(1)	-	-	$\frac{N(F)^{\times}}{\langle \overline{F}^{\times 2},\kappa_{u-1}+1\rangle}$	-	$\frac{(N(F)+N(F))^{\times}}{N(F)^{\times}}$
.42	$P_{u-2,1,0,2}$	Z_2	ιy_u	$F_{u-1,1,0,2}$	$\frac{N(F)^{\times}}{\overline{F}^{\times 2}}$	-	$\frac{\overline{F}^{\times}}{N(F)^{\times}}$
.41 & .42	$P_{u-1,1,0,2}$	(1)	-	-	$\frac{N(F)^{\times}}{\langle \overline{F}^{\times 2},\alpha_{u-1}\rangle}$	-	$\frac{\overline{F}^{\times}}{N(F)^{\times}}$
.511	P_{2102}	(1)	-	-	$\frac{N(F)^{\times}\cap(\overline{F}^2+\overline{F}^2)^{\times}}{\langle \overline{F}^{\times 2},2\rangle}$	-	$\frac{(N(F)+N(F))^{\times}}{N(F)^{\times}}$
.5211 & .5311	$P_{w-1,1,0,2}$	Z_2	$\frac{\gamma\kappa_3}{\kappa_{w+1}}\, y_{w+1}{}^{2}$	P_{w102}	$\frac{N(F)^{\times}\cap(\overline{F}^2+\overline{F}^2)^{\times}}{\overline{F}^{\times 2}}$	-	$\frac{(N(F)+N(F))^{\times}}{N(F)^{\times}}$
.5212 & .5312	$P_{w-1,1,0,2}$	(1)	-	-	$\frac{N(F)^{\times}\cap(\overline{F}^2+\overline{F}^2)^{\times}}{\overline{F}^{\times 2}}$	$(1/\kappa_{w+1})y_{w+1}$	$\frac{(N(F)+N(F))^{\times}}{\langle N(F)^{\times},\kappa_w+1\rangle}$
.5221 & .5321	$P_{w-1,1,0,2}$	(1)	-	-	$\frac{N(F)^{\times}\cap(\overline{F}^2+\overline{F}^2)^{\times}}{\langle \overline{F}^{\times 2},-1\rangle}$	-	$\frac{(N(F)+N(F))^{\times}}{N(F)^{\times}}$
.5211 & .5311	P_{w102}	(1)	-	-	$\frac{N(F)^{\times}\cap(\overline{F}^2+\overline{F}^2)^{\times}}{\langle \overline{F}^{\times 2},\kappa_w+1\rangle}$	-	$\frac{(N(F)+N(F))^{\times}}{N(F)^{\times}}$
.53 & .54	P_{k102} $(2 \le k < w-2)$	Z_2	y_{k+2}	$P_{k+1,1,0,2}$	$\frac{N(F)^{\times}\cap(\overline{F}^2+\overline{F}^2)^{\times}}{\overline{F}^{\times 2}}$	-	$\frac{(N(F)+N(F))^{\times}}{N(F)^{\times}}$
.531	$P_{w-2,1,0,2}$	Z_2	y_w	$P_{w-1,1,0,2}$	$\frac{N(F)^{\times}\cap(\overline{F}^2+\overline{F}^2)^{\times}}{\overline{F}^{\times 2}}$	-	$\frac{(N(F)+N(F))^{\times}}{N(F)^{\times}}$

Table 16 (continued)

Case	Group	OSA	OSA Gens	2-overgroups	sim. set	N Gens	Conjugacy of $Q_{k+1}(\gamma, \delta)$ set
.5321	$P_{w-2,1,0,2}$	Z_2	$\iota\delta y_w$ [3]	$P_{w-1,1,0,2}$	$\frac{N(F)^{\times}\cap(\overline{F}^2+\overline{F}^2)^{\times}}{\overline{F}^{\times 2}}$	–	$\frac{(N(F)+N(F))^{\times}}{N(F)^{\times}}$
.5322	$P_{w-2,1,0,2}$	(1)	–	–	$\frac{N(F)^{\times}\cap(\overline{F}^2+\overline{F}^2)^{\times}}{\overline{F}^{\times 2}}$	ιy_w	$\frac{(N(F)+N(F))^{\times}}{\langle N(F)^{\times},-1\rangle}$

[1] $N(\beta) = 1/\alpha_u$.

[2] $N(\gamma) = \kappa_w + 1$.

[3] $N(\delta) = -1$.

are similar to groups in $F^{\times} . U$. It can be shown that such a group in $F^{\times} . U$ is similar to some $Q_k(\gamma, \delta)$ via an element of V and where $x(\gamma, \delta) \in U$. The similarity invariant (and the conjugacy-by-elements-of-V invariant) of $Q_k(\gamma, \delta)$ is $-\det x(\gamma, \delta) = \gamma^2 + \delta^2$ (modulo the determinants of elements from the abelian maximal subgroup of $Q_k(\gamma, \delta)$). Now $x(\gamma, \delta) = \begin{pmatrix} \gamma & \delta \\ \delta & -\gamma \end{pmatrix} \in U$ iff $\gamma^2 + \delta^2 = \gamma/\overline{\gamma} = \delta/\overline{\delta}$. Set $N^{-1}(1) = \{\lambda \in F^{\times} \mid \exists\gamma, \delta \in F^{\times}, \lambda = \gamma^2+\delta^2 = \gamma/\overline{\gamma} = \delta/\overline{\delta}\}$, and then $N^{-1}(1)$ is a subgroup of $F^{\times}$. The scalar matrices in U have determinants from $N^{-1}(1)^2$ and so we are interested in $N^{-1}(1)/N^{-1}(1)^2$. The isomorphism γ , of 15.9 gives $N^{-1}(1)/N^{-1}(1)^2 \approx \left(N(F)^{\times} \cap \left(\overline{F}^2+\overline{F}^2\right)^{\times}\right)/\overline{F}^{\times 2}$. To obtain the set of possible similarity classes (*similarity set*) of the $Q_k(\gamma, \delta)$ in $F^{\times} . U$, this last must be adjusted slightly further by determinants of other elements in the abelian maximal subgroup of $Q_k(\gamma, \delta)$. All copies of a given $Q_k(\gamma, \delta)$ in $F^{\times} . U$ are conjugate by elements of V by 15.6 (i) and so for *each* similarity class of $Q_k(\gamma, \delta)$ in $F^{\times} . U$, we have conjugacy classes in $F^{\times} . U$ in one-to-one correspondence with cosets of $N_V(Q_k) . U$ in V . Once again we employ the isomorphism $\xi : V/F^{\times} . U \approx \left(N(F)+N(F)\right)^{\times}/N(F)^{\times}$ of 15.8 to obtain the conjugacy set for each similarity class of $Q_k(\gamma, \delta)$ and note the generators of $N_V(Q_k) . U/F^{\times} . U = N$ in terms of matrices in V . The conjugacy classes of a given P_{k10} $(k \geq 2)$ in PU are then in one-to-one correspondence with the cartesian product of the similarity set of $Q_{k+1}(\gamma, \delta)$ and the conjugacy set of the $Q_{k+1}(\gamma, \delta)$.

It should be noted that $N^{-1}(1) = N^{-1}(1)$ or again $N(F)^{\times} \leq \overline{F}^2 + \overline{F}^2$ whenever $u = u(F) \geq 2$ and this allows the notation to be simplified at times; this holds when $|F| < \infty$.

We write $\mathrm{OSA}(P_{k102}) = N_{\mathrm{PU}}(P_{k102})/P_{k102} \approx N_{F^{\times}.U}(Q_{k+1}(\gamma, \delta))/Q_{k+1}(\gamma, \delta)$.

18. PSU(2, F)

We seek those $Q_k(\gamma, \delta)$ which lie in $F^{\times} . SU$. In particular we must have that $\det x(\gamma, \delta) = -\gamma^2 - \delta^2 = 1$; that is, $-1 = \gamma^2 + \delta^2 = \gamma/\overline{\gamma} = \delta/\overline{\delta}$. This is equivalent to $\overline{s}(F) = 2$ and only one similarity class of Q_k's is involved. All

Table 17

Case	Group	OSA	OSA Gens	2-Overgroups	0 gens	Conjugacy Set
.1 & 2	P_{k102} $(2 \le k < u-1)$	Z_2	y_{k+2}	$P_{k+1,1,0,2}$	-	$\dfrac{(N(F)+N(F))^\times}{\overline{F}^{\times 2}}$
.12 & .22	$P_{u-1,1,0,2}$	(1)	-	-	$(1/\kappa_{u+1})y_{u+1}$	$\dfrac{(N(F)+N(F))^\times}{\langle \overline{F}^{\times 2}, \kappa_u+1\rangle}$
.3 & .4	P_{k102} $(2 \le k < u-2)$	Z_2	y_{k+2}	$P_{k+1,1,0,2}$	-	$\dfrac{(N(F)+N(F))^\times}{\overline{F}^{\times 2}}$
.32 & .42	$P_{u-2,1,0,2}$	(1)	-	-	y_u	$\dfrac{(N(F)+N(F))^\times}{\langle \overline{F}^{\times 2}, \kappa_{u-1}+1\rangle}$
.521 & .531	$P_{w-1,1,0,2}$ $(\overline{s}(F) = 2)$	(1)	-	-	$(1/\kappa_{w+1})y_{w+1}$	$\dfrac{(N(F)+N(F))^\times}{\langle \overline{F}^{\times 2}, \kappa_w+1\rangle}$
.53 & .54	P_{k102} $(2 \le k < w-2, \overline{s}(F) = 2)$	Z_2	y_{k+2}	$P_{k+1,1,0,2}$	-	$\dfrac{(N(F)+N(F))^\times}{\overline{F}^{\times 2}}$
.531	$P_{w-2,1,0,2}$ $(\overline{s}(F) = 2)$	Z_2	y_w	$P_{w-1,1,0,2}$	-	$\dfrac{(N(F)+N(F))^\times}{\overline{F}^{\times 2}}$
.532	$P_{w-2,1,0,2}$ $(\overline{s}(F) = 2)$	(1)	-	-	ιy_w	$\dfrac{(N(F)+N(F))^\times}{\langle \overline{F}^{\times 2}, -1\rangle}$

copies of the Q_k's are then conjugate under the action of V. The conjugacy classes of P_{k102} in PSU are in one-to-one correspondence with the cosets of $N_V(Q_{k+1})$. SU in V. To describe the conjugacy set, we use the isomorphism η described in 15.9:

$$V/F^\times \cdot \mathrm{SU} \approx (N(F)+N(F))^\times/\overline{F}^{\times 2} ,$$

and note generators of $0 = N_V(Q_{k+1}) \cdot \mathrm{SU}/F^\times \cdot \mathrm{SU}$ in terms of corresponding matrices in V. We set $\mathrm{OSA}(P_{k102}) = N_{\mathrm{PSU}}(P_{k102})/P_{k102}$ $(\approx N_{F^\times \cdot \mathrm{SU}}(Q_{k+1})/Q_{k+1})$ and give its generators as matrices in SU. Note also that if $u(\overline{F}) = 2$, then $(N(F)+N(F))^\times = \overline{F}^\times$, as in cases .1 and .4. As always, the conjugacy class of an overgroup is obtained by passing to the quotient in the appropriate conjugacy set.

REFERENCE

S.B. Conlon (1975), "p-groups with an abelian maximal subgroup and cyclic center", *J. Austral. Math. Soc.* (to appear).

Department of Pure Mathematics,
University of Sydney,
Sydney,
New South Wales, Australia.

PROC. MINICONF. THEORY OF GROUPS 20D10
CANBERRA 1975, 51-52.

A NOTE ON GROUPS WITH THE INVERSE LAGRANGE PROPERTY

T.M. Gagen

A group G has the inverse Lagrange property if for every divisor d of $|G|$, there exists a subgroup of order d. By the well-known theorem of P. Hall, any group with this property must be solvable but, of course, not every solvable group has a subgroup of every possible order. A quick induction shows that supersolvable groups satisfy the inverse Lagrange property. Certain non-supersolvable groups, for example S_4, have subgroups of every order.

Any solvable group G can be embedded in a group with the inverse Lagrange property. For if H is a cyclic group of order $|G|$, then $G \times H$ has subgroups of every order. The purpose of this note is to show that the true situation is more complicated than this. We show that every solvable group can be embedded in a directly indecomposable group with the inverse Lagrange property. Below is an outline of the construction. Let $G = G_1$ be some given solvable group.

1. Embed G_1 in a solvable directly indecomposable group G_2 which has a normal Hall subgroup N of prime index p. (This can be achieved by choosing a suitable subgroup of $G_1 \text{ wr } Z_p$ where p is a prime and $(p, |G|) = 1$, for example.) Let $|G_2| = pn$ and let P be a Sylow p-subgroup of G_2.

2. Let H be a cyclic group of order n and $G_3 = G_2 \times H$. Then G_3 has the inverse Lagrange property.

For let $d = p^{\varepsilon}p_1^{a_1} \dots p_k^{a_k}$ be a divisor of $|G_3|$. It is easy to see that the only interesting case occurs when $\varepsilon = 1$ and also if $p_1^{a_1} \dots p_k^{a_k} \mid |G_1| = |H|$, then H has a subgroup of this order. Thus we may assume that certain prime divisors of d occur to a power greater than that in $|G_2|$. Suppose without loss that these are $p_1, \ldots, p_l$. Write $d = p\, p_1^{b_1} \ldots p_l^{b_l} q$ where $p_i^{b_i} = |G_2|_{p_i}$, for $i = 1, \ldots, l$ and $q \mid |H|$. Choose a Hall subgroup H_1 of G_2 of order $p\, p_1^{b_1} \ldots p_l^{b_l}$ and a subgroup H_2 of H of order q . Then $H_1 \times H_2$ is a subgroup of order d .

3. Choose a prime r such that $r \equiv 1 \left(\bmod |G_2|\right)$. Such a prime exists by Dirichlet's Theorem. Let $G_4 = QG_3$ be the split extension of a cyclic group Q of order r by G_3 with G_3 acting on Q with kernel $N \subseteq G_2 \subseteq G_3$. Note that $N \trianglelefteq G_3$ and G_3/N is cyclic of order $|G_2|$.

The group G_4 is easily seen to have the inverse Lagrange property. Suppose $G_4 = A_1 \times \ldots \times A_s$, where A_i , for $i = 1, \ldots, s$, are indecomposable. Without loss $Q \subseteq A_1$ and since P is a Sylow subgroup of G_4 which acts non-trivially on Q , $P \subseteq A_1$. Since $A_2, \ldots, A_s \subseteq C_{G_4}(Q) = N \subseteq G_2$, by Dedekind's law, $G_2 = \left(G_2 \cap A_1\right) \times A_2 \times \ldots \times A_s$. But G_2 is indecomposable and $P \subseteq G_2 \cap A_1 \neq 1$. Thus $A_2 = \ldots = A_s = 1$ and $G_4 = A_1$ is indecomposable.

Department of Pure Mathematics,
University of Sydney,
Sydney,
New South Wales, Australia.

PROC. MINICONF. THEORY OF GROUPS 20H15
CANBERRA 1975, 53-56.

A NOTE CONCERNING COXETER GROUPS AND PERMUTATIONS

Marcel Herzog and Gustav I. Lehrer

Ree's theorem on permutations is generalized to Coxeter groups.

Let V be a Euclidean space of dimension n and suppose W is a Coxeter group generated by reflections in V (see [1, 2] for the precise definition). For $\alpha \in V$, we write H_α for the hyperplane orthogonal to α and r_α for reflection in H_α. For any $g \in O(V)$, the orthogonal group of V, denote by $n(g)$ and $k(g)$ the dimensions of $\mathrm{Im}(g-1)$ and $\mathrm{Ker}(g-1)$ respectively. If G is any subgroup of $O(V)$, let $(V, 1)_G$ denote the dimension of the subspace of G-fixed vectors in V, and write $r(G)$ for $n - (V, 1)_G$.

The object of this note is to prove

THEOREM A. *Let W be a Coxeter group generated by reflections in V and suppose $w_1, w_2, \ldots, w_m \in W$. Let G be the group generated by $w_1, \ldots, w_m$. Then*

$$n(w_1) + n(w_2) + \ldots + n(w_m) + n(w_1 \cdot \ldots \cdot w_m) \geq 2r(G) . \tag{1}$$

This is a generalization of Ree's result [5], a direct proof of which was given by Feit, Lyndon and Scott in [4].

For the proof we need the following simple results.

LEMMA 1 (Carter [3], Lemmas 2 and 3). *(i) Let $w \in W$ and denote by $m(w)$ the minimum length of a word expressing w as a product of reflections in W. Then $m(w) = n(w)$.*

(ii) Let $w = r_{\alpha_1} \cdot \ldots \cdot r_{\alpha_k}$, $r_{\alpha_i} \in W$. *Then* $k = n(w)$ *if and only if the* α_i *are linearly independent.*

COROLLARY 1'. *Let* $g \in W$ *and suppose* r_α *is a reflection in* W. *Then*

$$n(gr_\alpha) = \begin{cases} n(g) + 1 & \text{if } H_\alpha \not\supset \mathrm{Ker}(g-1), \\ n(g) - 1 & \text{if } H_\alpha \supset \mathrm{Ker}(g-1). \end{cases}$$

Proof. Let $g = r_{\alpha_1} \cdot \ldots \cdot r_{\alpha_k}$ be a reduced expression (that is $k = n(g)$). Then by Lemma 1, (ii), $\mathrm{Ker}(g-1) = H_{\alpha_1} \cap \ldots \cap H_{\alpha_k}$. If α is not a linear combination of $\alpha_1, \ldots, \alpha_k$ (that is $H_\alpha \not\supset H_{\alpha_1} \cap \ldots \cap H_{\alpha_k}$) then again by Lemma 1, (ii), the expression $gr_\alpha = r_{\alpha_1} \cdot \ldots \cdot r_{\alpha_k} r_\alpha$ is reduced and $n(gr_\alpha) = n(g) + 1$. If on the other hand, $H_\alpha \supset H_{\alpha_1} \cap \ldots \cap H_{\alpha_k}$ then $n(gr_\alpha) < k + 1$ and taking determinants we see that $n(gr_\alpha) = k - 1 = n(g) - 1$.

We next show that Theorem A is implied by

PROPOSITION B. *Let* $r_1, r_2, \ldots, r_m$ *be reflections in* W *and let* H *be the group generated by* $r_1, \ldots, r_m$. *Then*

$$m + n(r_1 \cdot \ldots \cdot r_m) \geq 2r(H) . \tag{2}$$

Note that this is the special case of Theorem A arising when the w_i are reflections.

Proof that Proposition B implies Theorem A. With notation as in the statement, let $w_i = r_{i1} \cdot \ldots \cdot r_{in_i}$ be reduced expressions for the w_i, $i = 1, \ldots, m$. Then $n(w_i) = n_i$ and $(V, 1)_G = \dim \bigcap_i \mathrm{Ker}(w_i-1) = \dim \bigcap_{i,j} \mathrm{Ker}(r_{ij}-1) = (V, 1)_H$ where H is the group generated by the r_{ij}. Thus $r(H) = r(G)$. From Proposition B we therefore have

$$n_1 + \ldots + n_m + n(w_1 \cdot \ldots \cdot w_m) \geq 2r(H) = 2r(G) .$$

which is the required statement.

Proof of Proposition B. This is by induction on $\dim V$. If $\dim V = 1$, there is only one reflection and the statement is trivial. Thus we assume $\dim V = n > 1$. Suppose $V = V_1 \perp V_2$, where the V_i are H-invariant. Then each r_i fixes either V_1 or V_2 pointwise and if r_i fixes V_1, r_j fixes V_2 then $r_i r_j = r_j r_i$

since V_1 is orthogonal to V_2. Thus (changing the indices if necessary) we can assume $r_1, r_2, \ldots, r_u$ fix V_1 and $r_{u+1}, \ldots, r_{u+v}$ fix V_2. Writing $H_2 = \langle r_1, \ldots, r_u \rangle$ and $H_1 = \langle r_{u+1}, \ldots, r_{u+v} \rangle$ we then have $H = H_1 \times H_2$ and $(V, 1)_H = (V_1, 1)_{H_1} + (V_2, 1)_{H_2}$. Thus $r(H) = r(H_1) + r(H_2)$ and if the V_i are proper subspaces of V the result follows by induction.

Thus we may assume that V has no H-invariant subspaces. Let $\alpha_{i_1}, \ldots, \alpha_{i_k}$ be a maximal linearly independent subset of the roots corresponding to the r_i (that is $r_i = r_{\alpha_i}$). Then the subspace of V spanned by the α_i is H-invariant, whence $k = n = \dim V$. Suppose $n(r_1 \cdot \ldots \cdot r_m) = n - k$ $(k \geq 0)$. Since there are at least n indices i such that

$$n(r_1 \cdot \ldots \cdot r_{i-1}r_i) = n(r_1 \cdot \ldots \cdot r_{i-1}) + 1$$

(by Corollary 1' and the above observation), there are at least k indices j such that $n(r_1 \cdot \ldots \cdot r_{j-1}r_j) = n(r_1 \cdot \ldots \cdot r_{j-1}) - 1$. As these two sets of indices are disjoint, we have $m \geq n + k$. Hence

$$m + n(r_1 \cdot \ldots \cdot r_m) \geq n + k + n - k = 2n = 2r(H)$$

which completes the proof.

COROLLARY A'. *If $w_1, \ldots, w_m$ generate a subgroup of W which fixes no vector of V, then*

$$n(w_1) + \ldots + n(w_m) \geq n \quad (= \dim V) . \tag{3}$$

If equality holds, then for the product w of the w_i in any order we have $n(w) = n$.

Let S_n be the symmetric group on $\{1, 2, \ldots, n\}$, regarded as a Coxeter group as usual (that is if $e_1, \ldots, e_n$ is an orthogonal basis of V, S_n is generated by the reflections r_{ij} in $H_{ij} : e_i - e_j = 0$). If $\pi \in S_n$ is a product of disjoint cycles of length $n_1, \ldots, n_r$ then clearly $n(\pi) = \sum_{i=1}^{r} (n_i - 1)$ (= $\nu(\pi)$ in Ree's notation). If G is a subgroup of S_n, then $(V, 1)_G$ is the number of orbits of G on $\{1, \ldots, n\}$.

Hence from Theorem A we have

COROLLARY A". *Suppose $\pi_1, \ldots, \pi_m$ generate a subgroup of S_n which has t*

orbits on $\{1, \ldots, n\}$. *Then*

(4) $$v(\pi_1) + \ldots + v(\pi_m) + v(\pi_1 \cdot \ldots \cdot \pi_m) \geq 2(n-t) .$$

This specializes to Ree's theorem when $\pi_1 \cdot \ldots \cdot \pi_m = 1$.

COROLLARY A"' (Ree). *If* $\pi_1, \ldots, \pi_m \in S_n$ *generate a transitive subgroup of* S_n *then*

$$v(\pi_1) + \ldots + v(\pi_m) \geq n - 1 .$$

If we have equality above, then the product of the π_i *in any order is an* n*-cycle.*

The latter statement follows because $v(\pi) = n - 1$ implies that π is an $(n-1)$-cycle (that is a Coxeter element). However for an arbitrary Coxeter group, equality in (3) does not imply that the products of the w_i in any order even have the same eigenvalues. For example let $\{e_1, e_2, e_3, e_4\}$ be an orthonormal basis of Euclidean 4-space and let D_4 have the usual root system (see [1, 2]). Let $w_1 = r_{e_1-e_2}$, $w_2 = r_{e_2-e_3}$, $w_3 = r_{e_1+e_2}$ and $w_4 = r_{e_4-e_1}$. Then equality holds in (3), but $w_1w_2w_3w_4$ and $w_1w_3w_2w_4$ are not conjugate in D_4 . In fact the former has characteristic polynomial $(t+1)(t^3+1)$ while the latter has characteristic polynomial $(t^2+1)^2$.

References

[1] N. Bourbaki, *Éléments de Mathématique*, Fasc. 34. *Groupes et Algèbres de Lie*, Chapitres 4, 5 et 6 (Actualités Scientifiques et Industrielles, 1337. Hermann, Paris, 1968). MR39#1590.

[2] N. Bourbaki, *Éléments de Mathématique*, Fasc. 37. *Groupes et Algèbres de Lie*, Chapitres 2 et 3 (Actualités Scientifiques et Industrielles, 1349. Hermann, Paris, 1972). Zb1M244#22007.

[3] R.W. Carter, "Conjugacy classes in the Weyl group", *Compositio Math.* 25 (1972), 1-59. MR47#6884.

[4] Walter Feit, Roger Lyndon and Leonard L. Scott, "A remark about permutations", *J. Combinatorial Theory Ser. A* 18 (1975), 234-235. Zb1M297#05021.

[5] Rimhak Ree, "A theorem on permutations", *J. Combinatorial Theory Ser. A* 10 (1971), 174-175. MR42#4414.

Department of Mathematics, IAS,
Australian National University,
Canberra, ACT, Australia.

Department of Pure Mathematics,
University of Sydney,
Sydney, New South Wales, Australia.

PROC. MINICONF. THEORY OF GROUPS
CANBERRA 1975, 57-65.

20K99

STRONGLY PURE SUBGROUPS OF ABELIAN GROUPS

S. Janakiraman and K.M. Rangaswamy

All the groups considered here are abelian and the reader is referred to [2, 3] for general results, notation and terminology used in this paper.

The present work arose out of the observation that a subgroup S of an abelian p-group G without elements of infinite height is pure in G if and only if to each $a \in S$, there exists a homomorphism $\alpha : G \to S$ satisfying $\alpha(a) = a$. In general, the latter property turns out to be stronger than purity (see Lemma 3). Accordingly, we call a subgroup S of an arbitrary abelian group G *strongly pure* if to each $s \in S$, there exists a homomorphism $f : G \to S$ such that $f(s) = s$.

Strong purity has been found useful in module theory. For example, Villamayor has shown (see [1]) that a module A over a ring R is flat exactly when $A \cong P/S$, where P is projective and the submodule S is strongly pure in our sense. Also, modules which are strongly pure in every containing module are shown in [4] to be precisely the finitely injective modules.

In this article, strong purity is analysed in some detail. Strongly pure injective and strongly pure projective abelian groups are characterised. The notion of the strongly pure direct limit is introduced and is used to study the groups G with the property that G is strongly pure in every group in which G is a pure subgroup. Specifically, a torsion or torsion-free group G will have the last mentioned property if and only if G is a strongly pure direct limit of algebraically compact abelian groups. Finally, we also investigate the abelian groups in which pure subgroups are strongly pure.

LEMMA 1. *Let G be a p-group. A subgroup S is strongly pure in G if and only if, to each $a \in S[p]$, there exists $f : G \to S$, with $f(a) = a$.*

Proof. By induction on the order of the element. Assume that to each element of order less than or equal to p^n in S, there is a homomorphism from G to S fixing that element. Let s be an element of order p^{n+1} in S. Let $f : G \to S$ satisfy $f(ps) = ps$. Since $s - f(s) \in S[p]$, there exists $g : G \to S$ satisfying $g(s-f(s)) = s - f(s)$. Let $h = f + g - gf$. Then h is a homomorphism from G to S satisfying $h(s) = s$. Thus S is strongly pure.

LEMMA 2. *If S is strongly pure in G, then, for any finite set of elements $s_1, \ldots, s_n$, there exists a $\theta : G \to S$ with $\theta(s_i) = s_i$, $i = 1, \ldots, n$.*

Proof. As before we apply induction on n. Let $n = k + 1$ and assume that the result is true for $n = k$. Let $f : G \to S$ with $f(s_i) = s_i$, $i = 1, \ldots, k$. Let $g : G \to S$ satisfy $g(s_{k+1}-f(s_{k+1})) = s_{k+1} - f(s_{k+1})$. If $h = g + f - gf$, then $h(s_i) = s_i$ for $i = 1, \ldots, k+1$. Hence the result.

It is easy to see that strong purity is transitive, that is, if $A \subset B \subset G$ with A strongly pure in B and B strongly pure in G, then A is strongly pure in G. Also, if A is strongly pure in G, then A is strongly pure in every subgroup between A and G. It is clear from Lemma 2 that a finitely generated strongly pure subgroup is a direct summand.

LEMMA 3. *A strongly pure subgroup of a group G is isotype in G and hence, in particular, pure. But the converse does not hold.*

Proof. Let S be strongly pure in G. It is enough to show that $S \cap p^\alpha G \subseteq p^\alpha S$ for every prime p and ordinal α. Let $s \in S \cap p^\alpha G$. By hypothesis, there exists $f : G \to S$ with $f(s) = s$. Since a homomorphism does not decrease heights, $s \in p^\alpha S$. Thus S is isotype in G.

An isotype (hence, a pure) subgroup need not be strongly pure. For example, let G be an indecomposable torsion-free abelian group of finite rank. Let S be a proper non-zero pure subgroup. Then S is not strongly pure in G. For, otherwise, corresponding to a maximal independent subset $\{s_1, \ldots, s_n\}$ of S there will be a $\theta : G \to S$ satisfying $\theta(s_i) = s_i$, $i = 1, \ldots, n$. This θ will act as identity on S making S a direct summand, contradicting the indecomposability of G. Another useful example is the additive group J_p of p-adic integers. It is readily seen that 0, J_p are the only strongly pure subgroups while J_p has 2^c distinct pure ($\equiv$ isotype) subgroups, where c is the cardinality of the continuum.

However, under certain restrictions, purity implies strong purity.

PROPOSITION 4. *Let G be a p-group and S a subgroup with $S \cap p^\omega G = 0$.*

Then S *is pure if and only if* S *is strongly pure.*

Proof. Let S be pure with $p^{\omega}S = S \cap p^{\omega}G = 0$. Let $0 \neq s \in S$. Since s has finite height, s can be imbedded in a finitely generated summand D of S . Since D is a summand of G , we have a projection $\eta : G \to D$ which satisfies $\eta(s) = s$. Thus S is strongly pure in G .

COROLLARY 5. *Let* G *be a* p*-group without elements of infinite height. Then all the pure subgroups of* G *are strongly pure.*

PROPOSITION 6. *Let* G *be a totally projective* p*-group. Then a subgroup* S *is strongly pure exactly when* S *is isotype.*

Proof. Let S be isotype in G and $s \in S[p]$. Clearly the inclusion map $\theta : \langle s \rangle \to S$ is height non-decreasing. Since $\langle a \rangle$ is nice in G and $G/\langle a \rangle$ is totally projective, we appeal to Corollary 81.4 of [3] to extend θ to a homomorphism $\eta : G \to S$. Thus S is strongly pure in G .

PROPOSITION 7. *Let* F *be a free abelian group (or more generally, the reduced part of* F *is a direct sum of copies of a fixed subgroup of rational numbers). Then every pure subgroup of* F *is strongly pure.*

Proof. One has just to note that every pure subgroup of finite rank in F is a direct summand (see Lemma 86.8 in [3]).

Strong purity does not in general seem to be as well behaved as purity. For example, if A is strongly pure in G and $S \leq A$, then A/S need not be strongly pure in G/S . To see this, let F be an uncountable free abelian group and S a (pure) subgroup such that F/S is the additive group of p-adic integers. Let $S \lneqq A \lneqq F$ with A pure in F . By Proposition 7, A is strongly pure in F , but as noted in Lemma 3, A/S is not strongly pure in F/S . Also if $A \leq B \leq G$ with A strongly pure in G and B/A strongly pure in G/A , then B need not be strongly pure in G . For example, let $G = \prod_{p \in P} Z/pZ$ and $A = \bigoplus_{p \in P} Z/pZ$, where P is the set of all primes. Let e_p be a generator of Z/pZ . Let B be the pure subgroup generated by A and $e = \langle \ldots, e_p, \ldots \rangle$ so that B/A is a direct summand of the divisible group G/A . Thus A and B/A are strongly pure respectively in G and C/A . But B is not strongly pure in G since any endomorphism α of G satisfying $\alpha(e) = e$ must be the identity on G .

Call an abelian group G strongly pure injective if given any row exact diagram

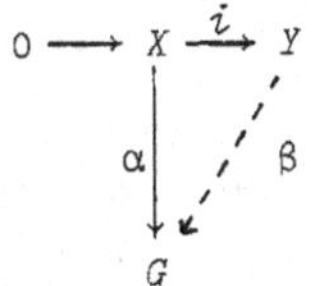

where $i(X)$ is strongly pure in Y, there exists $\beta : Y \to G$ satisfying $\beta i = \alpha$.

PROPOSITION 9. *An abelian group G is strongly pure injective if and only if G is algebraically compact.*

Proof. Let G be strongly pure injective. Consider an exact sequence $0 \to K \xrightarrow{i} F \to Q \to 0$, where Q is the additive group of rational numbers and F is a free abelian group. Now $i(K)$ is pure in F and hence, by Proposition 7, is strongly pure. Applying the functor Hom(, G), we get an exact sequence

$$\mathrm{Hom}(F, G) \xrightarrow{i^*} \mathrm{Hom}(K, G) \to \mathrm{Ext}(Q, G) \to \mathrm{Ext}(F, G) = 0 .$$

Since G is strongly pure injective, i^* is epic so that $\mathrm{Ext}(Q, G) = 0$. This implies that G is cotorsion. We claim that G is algebraically compact. Without loss of generality we may assume that G is reduced. Let B be a basic subgroup of the p-component G_p, where p is an arbitrary prime relevant to G. Let $i : B \to \overline{B}$ be the imbedding map of B into its torsion completion $\overline{B}$. Now $i(B)$ is pure and hence, by Proposition 4, strongly pure in $\overline{B}$. Then the inclusion map $j : B \to G$ extends to an $\alpha : \overline{B} \to G$ so that $\alpha i = j$. It is clear that α is monic and that $\alpha(\overline{B}) \subseteq G_p$. Since B is pure in G and $\overline{B}/B$ is divisible, $\alpha(\overline{B}) \cong \overline{B}$ is pure in G_p and hence a summand of G_p. Since G is reduced, $G_p = \alpha(\overline{B}) \cong \overline{B}$ is torsion-complete, for all primes p. Then, by Theorem 1 of [5], we conclude that G is algebraically compact.

The converse follows from the fact that an algebraically compact abelian group is pure injective and hence is strongly pure injective.

One natural question that arises is, whether there exist enough strongly pure injectives, namely, whether every abelian group can be imbedded as a strongly pure subgroup of a strongly pure injective abelian group. The answer is 'no'. For example, the additive group Z of integers (more generally, any finite rank reduced torsion-free abelian group A) becomes a summand in any group G in which it is a strongly pure subgroup. Since Z (or A) is not algebraically compact, it cannot be imbedded as a strongly pure subgroup of a strongly pure injective abelian group.

Dually, one may define strongly pure projective abelian groups. These turn out to be just the pure projectives.

PROPOSITION 10. *An abelian group A is strongly pure projective exactly when A is a direct sum of cyclic groups. There exist enough strongly pure projective abelian groups.*

Proof. First observe that if G is a direct sum of cyclic groups, then any finitely generated pure subgroup of G is a direct summand so that any pure subgroup of G is strongly pure. The result then follows from the well known fact (see Lemma 30.19 [2]) that to every group A there exists an exact sequence

$0 \to K \xrightarrow{i} G \to A \to 0$ where G is a direct sum of cyclic groups and $i(K)$ pure in G.

We say a group G has the property (P), if G is strongly pure in any group containing G as a pure subgroup. Groups with property (P) turn out to be unions of algebraically compact groups.

LEMMA 11. *G has the property (P) if and only if G is a strongly pure subgroup of its pure injective hull $\hat{G}$.*

Proof. We need only to prove the sufficiency. Let G be pure in a group A. Then $\hat{G}$ is a pure subgroup of the pure injective hull $\hat{A}$ and hence a summand of $\hat{A}$. Since G is strongly pure in $\hat{G}$, it is strongly pure in $\hat{A}$ and hence in A.

It is easy to see that a group G has the property (P) if and only if its reduced part has the same property. Hence in the following discussion we may assume without loss of generality that G is reduced.

LEMMA 12. *If G is reduced and has the property (P), then $G^1 = \cap n!G = 0$.*

Proof. Now the pure-injective hull $\hat{G}$ of G is of the form $\hat{G} = D(G^1) \oplus E$, where $D(G^1)$ is the divisible hull of G^1 and E is reduced algebraically compact (see 41.8 of [2]). If $0 \neq g \in G^1$, by strong purity, there exists $\alpha : \hat{G} \to G$ with $\alpha(g) = g$. But then $0 \neq \alpha(D(G^1))$ is a divisible subgroup of G, contradicting the fact that G is reduced. Hence $G^1 = 0$.

PROPOSITION 13. *A reduced group G has the property (P) if and only if $G = \bigcup_{a \in G} S_a$, where S_a is algebraically compact, $a \in S_a$ and there is a homomorphism $\eta_a : G \to S_a$ satisfying $\eta_a(a) = a$.*

Proof. Suppose G has the property (P). By Lemma 12, $G^1 = 0$ so that the pure injective hull $\hat{G}$ of G is reduced algebraically compact. To each $a \in G$, there exists $\eta'_a : \hat{G} \to G$ with $\eta'_a(a) = a$. Let $S_a = \eta'_a(\hat{G})$. Since $G^1 = 0$, S_a is algebraically compact and we are done if we define η_a to be the restriction of η'_a to G.

The converse follows from Lemma 11 if one notes that G is pure in its pure injective hull $\hat{G}$ and that, by the pure injectivity of S_a, the map $\eta_a : \hat{G} \to S_a$ extends to $\eta'_a : \hat{G} \to S_a$.

Before we proceed further, we wish to introduce the following concept.

Let $\{A_i, i \in I; \pi_i^j\}$ be a direct system of abelian groups where I is a directed set and, for $i \leq j$, $\pi_i^j : A_i \to A_j$ are the morphisms of the direct system. The system $\{A_i, i \in I; \pi_i^j\}$ is called a *strongly pure direct system* if, for all $i, j \in I$ with $i \leq j$, $\pi_i^j(A_i)$ is strongly pure in A_j and the strong purity of $\pi_i^j(A_i)$ is compatible with the morphisms of the direct system: specifically, for each $i \in I$ and each $x_i \in A_i$, there exists, for each $j \geq i$, an $\eta_j : A_j \to \pi_i^j A_i$ such that $\eta_j \pi_i^j(x_i) = \pi_i^j(x_i)$ and, for any $k \geq j \geq i$, the diagram

$$\begin{array}{ccc} A_j & \xrightarrow{\pi_j^k} & A_k \\ \downarrow \eta_j & & \downarrow \eta_k \\ A_j & \xrightarrow{\pi_j^k} & A_k \end{array}$$

is commutative, that is, $\pi_j^k \eta_j = \eta_k \pi_j^k$.

If $\{A_i, i \in I; \pi_i^j, \eta_j\}$ is a strongly pure direct system and $A = \varinjlim A_i$, then it is readily seen that, for each i , $\pi_i(A_i)$ is strongly pure in A , where $\pi_i : A_i \to A$ is the canonical morphism of the direct limit. We call A the *strongly pure direct limit* of the groups A_i .

It is clear that if $G = \bigoplus_{k \in K} G_k$, then G can be obtained as the strongly pure direct limit of groups which are direct sums of finitely many G_k's .

It is well known that every group is a direct limit of finitely generated groups. Which groups are strongly pure direct limits of finitely generated abelian groups? This is answered below.

PROPOSITION 14. *The following properties are equivalent for an arbitrary abelian group* G *:*

(i) G *is a strongly pure direct limit of finitely generated abelian groups;*

(ii) $G^1 = 0$ *and* $G/T(G)$ *is* $\aleph_1$*-free and separable.*

Proof. Assume *(i)*. Let $G = \varinjlim \{G_i, i \in I; \pi_i^j, \eta_j\}$ be a strongly pure direct limit and let $\pi_i : G_i \to G$ be the canonical map. Since $\pi_i(G_i)$ is strongly pure and finitely generated, it is a direct summand of G . Thus every finitely generated subgroup of G can be imbedded in a finitely generated direct summand. It then follows that G has no non-zero elements of infinite height and that every pure subgroup of finite rank in $G/T(G)$ is a free summand, whence (see [2, 3]) $G/T(G)$ is $\aleph_1$-free and separable. This proves *(ii)*.

Assume *(ii)*. Let A be a finitely generated subgroup of G. Since $G/T(G)$ is $\aleph_1$-free and separable, $G/T(G) = B/T(G) \oplus C/T(G)$, where $B/T(G)$ is finitely generated free and $A \subset B$. Now $B = T(G) \oplus B'$, with $B' \cong B/T(G)$. If $\eta : B \to T(G)$ is the corresponding projection, then, since $G^1 = 0$, the finitely generated subgroup $\eta(A)$ can be imbedded in a finitely generated summand D (of $T(G)$ and hence) of G; $G = D \oplus E$. Clearly, $G = B' \oplus C = B' \oplus D \oplus (C \cap E)$ and $A \subseteq B' \oplus D$. Thus G is a directed union of finitely generated summands and this proves *(i)*.

Remark. One might analogously define a pure direct limit to be the direct limit of a direct system of groups where the images of the morphisms of the direct system are pure. It is easy to see that a group G is a pure direct limit of finitely generated groups exactly when $G^1 = 0$ and $G/T(G)$ is torsion-less, that is, $G/T(G)$ is a subgroup of a direct product of infinite cyclic groups.

PROPOSITION 15. *Let G be a reduced abelian group which is torsion or torsion-free. Then the following properties are equivalent for G:*

(i) G has the property (P);

(ii) $G^1 = 0$ and G is a strongly pure direct limit of reduced algebraically compact abelian groups;

(iii) G is a directed union of summands S_i, $i \in I$, where each S_i is reduced algebraically compact.

Proof. It is known that if $A^1 = 0$ and A is a homomorphic image of an algebraically compact group, then A itself is algebraically compact. From this it is clear that *(ii)* and *(iii)* are equivalent. Also, in view of Proposition 13, *(iii)* $\Rightarrow$ *(i)*.

Assume *(i)*. By Lemma 12, $G^1 = 0$. If G is torsion, then any finite subset of G can be imbedded in a finitely generated summand and hence G satisfies *(iii)*. Suppose G is torsion-free. Now G is strongly pure in its completion $\hat{G}$ under the n-adic topology, so that for each finite set $S = \{a_1, \ldots, a_n\}$ of G, there exists $\eta : \hat{G} \to G$ with $\eta(a_i) = a_i$, $i = 1, \ldots, n$. Because G is torsion-free, η acts as identity on the pure subgroup S^* generated by S. Since η is continuous, η is the identity on the completion C of S^* in the n-adic topology. Thus $G \supset C$. But the completion of a pure subgroup in $\hat{G}$ is a direct summand of $\hat{G}$. Thus G is a directed union of algebraically compact summands. This proves *(iii)*.

Remark. If an arbitrary mixed group G is a strongly pure direct limit of algebraically compact abelian groups, then it is clear that G has the property (P) . We are unable to say whether the converse holds. It is however easy to see that if G has the property (P) , then both $T(G)$ and $G/T(G)$ have the property (P) and hence are directed unions of algebraically compact direct summands.

Let us say that an abelian group G has the property (Q) if every pure subgroup of G is strongly pure. It is clear that if G has the property (Q) , then every pure subgroup also has the property (Q) . Also G has the property (Q) exactly when every countable pure subgroup is strongly pure.

PROPOSITION 16. *If G has the property (Q), then $G/T(G) = D \oplus E$, where D is divisible and E is a reduced homogeneous separable torsion-free abelian group and, for each prime p, either the p-component G_p is (divisible) $\oplus$ (bounded) or* $p^{\omega+1}G_p = 0$.

Proof. It is clear that $G/T(G)$ also has the property (Q) and hence every pure subgroup of finite rank in $G/T(G)$ is a direct summand. This is equivalent to saying (see [2, 3]) that $G/T(G)$ is a direct sum of a divisible group and a reduced homogeneous separable group. For an arbitrary prime p , consider the p-component G_p . If a basic subgroup of G_p is bounded, then G_p is (divisible) $\oplus$ (bounded) . Otherwise the final rank of a high subgroup of G_p is infinite. If $0 \neq a \in p^{\omega}G[p]$, then $\left|\left(p^{\omega+1}G \cap \langle a\rangle\right)/p\langle a\rangle\right| \leq p$ so that, by the main theorem in [6], $\langle a\rangle = p^{\omega}G_p \cap S$, for some pure subgroup S of G . By hypothesis, S is strongly pure in G and so $\langle a\rangle$ is pure in $p^{\omega}G_p$. This implies that the element a has zero p-height in $p^{\omega}G_p$, whence $p^{\omega+1}G_p = 0$.

References

[1] S.U. Chase, "Direct products of modules", *Trans. Amer. Math. Soc.* 97 (1960), 457-473. MR22#11017.

[2] László Fuchs, *Infinite abelian groups*, I (Pure and Applied Mathematics, 36. Academic Press, New York, London, 1970). MR41#333.

[3] László Fuchs, *Infinite abelian groups*, II (Pure and Applied Mathematics, 36-II. Academic Press, New York, London, 1973). MR50#2362.

[4] V.S. Ramamurthi and K.M. Rangaswamy, "On finitely injective modules", *J. Austral. Math. Soc.* **16** (1973), 239-248. MR48#11207.

[5] K.M. Rangaswamy, "A note on algebraic compact groups", *Bull. Acad. Polon. Sci. Ser. Sci. Math. Astronom. Phys.* **12** (1964), 369-371. MR29#5900.

[6] Fred Richman and Carol P. Walker, "On a certain purification problem for primary abelian groups", *Bull. Soc. Math. France* **94** (1966), 207-210. MR34#4360.

Department of Mathematics,
Madurai University,
Madurai,
India;

Department of Mathematics,
Australian National University,
Canberra, ACT, Australia;

Present Address:
Department of Mathematics,
University of Papua New Guinea,
Port Moresby,
Papua New Guinea.

PROC. MINICONF. THEORY OF GROUPS
CANBERRA 1975, 66-72. 20J99

RELATIVE COHOMOLOGY OF GROUPS

Hans Lausch

1. Introduction

This talk is to present a few results on cohomology of groups arising from an investigation into the cohomological behaviour of inverse semigroups ([6], [7], [8]). In this context, relative cohomology in the sense of Auslander ([1]) is just a special case of a more general situation, and various long exact sequences, some of which are well-known ([9], [11]) and have been established by various authors individually, can be obtained by one and the same method.

The starting point for our consideration is the concept of a *semilattice of groups*, that is a functor S from a semilattice E regarded as a category to the category Gp of groups. Thus for $e, f \in E$, $e \geq f$, there is exactly one group homomorphism $\phi_{e,f} : S(e) \to S(f)$ such that $\phi_{e,e}$ is the identity on $S(e)$, $e \in E$, and $e \geq f \geq g$ implies $\phi_{e,g} = \phi_{e,f}\phi_{f,g}$. It is convenient to regard S as a semigroup via $s_e s_f = (s_e\phi_{e,ef})(s_f\phi_{f,ef})$, for $s_e \in S(e)$, $s_f \in S(f)$, and $ef = \min(e, f)$.

Let E be a semilattice, $S : E \to \text{Gp}$ a semilattice of groups, $A : E \to \text{Gp}$ a semilattice of groups where $A(e)$, $e \in E$, are additively written abelian groups, and $(a, s) \to as$ a mapping from $A \times S \to A$ subject to the conditions:

(i) $(a_1+a_2)s = a_1s + a_2s$, $a_1, a_2 \in A$, $s \in S$;

(ii) $a(s_1s_2) = (as_1)s_2$, $a \in A$, $s_1, s_2 \in S$,

(iii) $a1_{S(e)} = a + 0_{A(e)}$, $a \in A$, $1_{S(e)}$ being the identity of $S(e)$, $0_{A(e)}$ being the zero of $A(e)$;

(iv) $0_{A(e)}s = 0_{A(ef)}$ if $s \in S(f)$.

A together with this map is called an *S-module*. Let A, B be two S-modules. A map $\alpha : A \to B$ will be called an *S-morphism* if

(i) $(a_1+a_2)\alpha = a_1\alpha + a_2\alpha$, $a_1, a_2 \in A$,

(ii) $(a\alpha)s = (as)\alpha$, $a \in A$, $s \in S$,

(iii) $A(e)\alpha \subseteq B(e)$, $e \in E$.

The set of all S-morphisms will be denoted by $\mathrm{Hom}_S(A, B)$ which becomes an abelian group via $a(\alpha+\beta) = a\alpha + a\beta$. S-modules together with S-morphisms then form an abelian category and one can show that this category has enough projectives and enough injectives ([6]). This allows us to define uniquely a cohomological functor $H_S = \left\{H_S^i \mid i \in Z\right\}$ from the category of S-modules to the category of abelian groups characterized by the following properties:

(i) $H_S^i(A) = 0$ if $i < 0$;

(ii) $H_S^i(J) = 0$ if $i > 0$ and J is injective;

(iii) $H_S^0(A) = \{\delta : E \to A \mid e\delta \in A(e), (e\delta)s = (ef)\delta$, for $s \in S(f)\}$ with $e(\delta_1+\delta_2) = e\delta_1 + e\delta_2$ as the group operation.

Note that for E being a singleton, H_S is just the ordinary cohomology of groups.

2. Change of semilattices

DEFINITION. Let $F \leq E$ be semilattices. We say F is *well-placed* in E if $Ee \cap F$ is a principal ideal in F for all $e \in E$; that is, there is a map $\psi : E \to F$ such that $Ee \cap F = F(e\psi)$.

PROPOSITION 2.1. *Let C be a convex subset of a semilattice F and F be well-placed in the semilattice E . Then $\Phi_E^F(C) = \{e \in E \mid e\psi \in C\}$ is a convex subset of E .*

Proof. Obvious.

DEFINITION. Let $S : E \to \underline{\underline{\mathrm{Gp}}}$ be a semilattice of groups and A an S-module such that $S(e \geq f) : A(e) \to A(f)$ is an isomorphism for all $e \geq f$ in E . Then A is called a homogeneous S-module.

DEFINITION. Let C be a convex subset of a semilattice E, $S : E \to \underline{\underline{\mathrm{Gp}}}$ a semilattice of groups and A a homogeneous S-module. Then (C, A) will denote the S-module defined by $(C, A)(e) = A(e)$ if $e \in C$, $(C, A)(e) = 0$ if $e \notin C$,

$(C, A)(e \geq f) = A(e \geq f)$ for $e, f \in C$, $(C, A)(e \geq f) = 0$ otherwise.

We state the following theorem without proof. Its proof can be found in [8]. It should be noted that it is a generalization of the theorem that states that a module A over a group obtained from coinduction of a module B over a subgroup is cohomologically equivalent to B.

THEOREM 2.2. *Let F be a well-placed subsemilattice of a semilattice E, $S : E \to \underline{\underline{Gp}}$ a semilattice of groups, $S_F : F \xrightarrow[\text{incl.}]{} E \xrightarrow[S]{} \underline{\underline{Gp}}$ its restriction to F, A a homogeneous S-module, $A_F : F \xrightarrow[\text{incl.}]{} E \xrightarrow[A]{} \underline{\underline{Gp}}$ its restriction to F, and C a convex subset of F. Then there is an isomorphism*

$$H^i_S\left(\Phi^F_E(C), A\right) \cong H^i_{S_F}\left(C, A_F\right), \text{ for all } i \in Z,$$

which is an isomorphism of δ-functors (see [5]).

Let $E = \{l, h, k, g \mid l \geq h, k \geq g, h, k$ incomparable$\}$. The following list describes $\Phi^F_E(C)$ for various well-placed subsemilattices F and convex subsets C of F:

C	F	$\Phi^F_E(C)$
g	g	g, h, k, l
h	h	h, l
k	k	k, l
l	l	l
g	g, h	g, k
g	g, k	g, h
g	g, l	g, h, k
h	h, l	h
k	k, l	k

3. Exact cohomology sequences

Let E be the semilattice as described at the end of §2, $S : E \to \underline{\underline{Gp}}$ a semilattice of groups and A a homogeneous S-module. We consider the following short exact sequences:

$$0 \to (g, A) \to (g, h, A) \to (h, A) \to 0 ,$$
$$0 \to (g, k, A) \to (g, h, k, A) \to (h, A) \to 0 ,$$
$$0 \to (h, k, A) \to (h, k, l, A) \to (l, A) \to 0 ,$$
$$0 \to (k, A) \to (h, k, l, A) \to (h, l, A) \to 0 ,$$
$$0 \to (g, A) \to (g, h, k, l, A) \to (h, k, l, A) \to 0 ,$$
$$0 \to (g, A) \to (g, h, k, A) \to (h, k, A) \to 0 ,$$
$$0 \to (h, k, l, A) \xrightarrow{\Delta} (h, l, A) \oplus (k, l, A) \to (l, A) \to 0 ,$$
$$0 \to (g, A) \xrightarrow{\Delta} (g, h, A) \oplus (g, k, A) \to (g, h, k, A) \to 0 ,$$
$$0 \to (g, h, k, A) \to (g, h, k, l, A) \to (l, A) \to 0 ,$$

where Δ means the diagonal map.

We observe that if $F = \{e\}$, $C = \{e\}$, for some $e \in E$, then $H^i_{S_F}(e, A) \cong H^i\big(S(e), A(e)\big)$, $i \in Z$, the ordinary cohomology of $S(e)$ with coefficients in $A(e)$. If $F = \{e, f \mid e \geq f\}$ and $C = \{f\}$, $e, f \in E$, we will write $H^i\big[S(f), S(e), A\big]$ for $H^i_{S_F}(f, A)$. Using Theorem 2.2 and the list at the end of §2, the cohomology functor H_S turns the short exact sequences above into long exact sequences as follows:

$$0 \to H^1_S(g, A) \to H^1\big[S(g), S(k), A\big] \to H^1\big[S(h), S(l), A\big] \to H^2(g, A) \to$$
$$\to H^2\big[S(g), S(k), A\big] \to \dots$$

$$0 \to H^1\big[S(g), S(h), A\big] \to H^1\big[S(g), S(l), A\big] \to H^1\big[S(h), S(l), A\big] \to$$
$$\to H^2\big[S(g), S(h), A\big] \to \dots$$

$$0 \to H^0_S(h, k, l, A) \to H^0\big[S(l), A(l)\big] \to H^1\big[S(h), S(l), A\big] \oplus H^1\big[S(k), S(l), A\big] \to$$
$$\to H^1_S(h, k, l, A) \to \dots$$

$$0 \to H^0_S(h, k, l, A) \to H^0\big[S(h), A(h)\big] \to H^1\big[S(k), S(l), A\big] \to H^1(h, k, l, A) \to \dots$$

$$0 \to H^0\big[S(g), A(g)\big] \to H^0_S(h, k, l, A) \to H^1_S(g, A) \to H^1\big[S(g), A(g)\big] \to$$
$$\to H^1_S(h, k, l, A) \to \dots$$

$$0 \to H^1_S(g, A) \to H^1\big[S(g), S(l), A\big] \to H^1\big[S(h), S(l), A\big] \oplus H^1\big[S(k), S(l), A\big] \to$$
$$\to H^2_S(g, A) \to \dots$$

$$0 \to H^0_S(h, k, l, A) \to H^0\big[S(h), A(h)\big] \oplus H^0\big[S(k), A(k)\big] \to H^0\big[S(l), A(l)\big] \to$$
$$\to H^1_S(h, k, l, A) \to \dots$$

$$0 \to H_S^1(g, A) \to H^1(S(g), S(h), A) \oplus H^1(S(g), S(k), A) \to H^1(S(g), S(l), A) \to$$
$$\to H_S^2(g, A) \to \dots$$

$$0 \to H^0(S(g), A(g)) \to H^0(S(l), A(l)) \to H^1(S(g), S(l), A) \to H^1(S(g), A(g)) \to \dots .$$

We first discuss the last sequence. Let N be a normal subgroup of $S(l)$ and $S(g) = S(l)/N$, $A(l)$ an $S(l)$-module $A(l)^N = H^0(N, A(l)) = A(g)$. Then the sequence becomes a special case of the Hochschild-Serre sequence, but of infinite length:

$$0 \to H^1(S(l)/N, A(l)) \to H^1(S(l), A(l)) \to H^1(S(l)/N, S(l), A)$$
$$\to H^2(S(l)/N, A(l)) \to H^2(S(l), A(l)) \to H^2(S(l)/N, S(l), A) \to \dots .$$

If $S(l)$ is a subgroup of $S(g)$, then the last sequence shows that $H^i(S(g), S(l), A)$ is just the relative cohomology in the sense of Auslander (see [1], also [9]).

Let us assume henceforward that $S(e \geq f)$ is a monomorphism whenever $f \leq e \in E$. It is then a consequence of a result by Ribes ([9]) that $H_S^1(g, A) = 0$ for all homogeneous S-modules A if and only if $S(g) = \langle S(h), S(k) \rangle$. Another result by the same author ([10]) implies that $H_S^i(g, A) = 0$ for all $i \in Z$ and all homogeneous S-modules A if and only if $S(g)$ is the free product of $S(h)$ by $S(k)$ with amalgamated subgroup $S(l)$. One of the long sequences above then implies that $H_S^i(h, k, l, A)$ is just the i-th cohomology of this amalgamated product with coefficients in $A(g)$. In this case we can state the following theorem which is just an interpretation of the sequences above.

THEOREM 3.1. *Suppose $S(g)$ is the free product of $S(h)$ and $S(k)$ with amalgamated subgroup $S(l)$. Then*

$$H^i(S(g), S(k), A) \cong H^i(S(h), S(l), A) , \quad i = 1, 2, \dots$$

(excision theorem)

$$H^i(S(g), S(l), A) \cong H^i(S(h), S(l), A) \oplus H^i(S(k), S(l), A) , \quad i = 1, 2, \dots$$

(coproduct theorem)

and there is the following exact sequence:

$$0 \to H^0(S(g), A(g)) \to H^0(S(h), A(h)) \oplus H^0(S(k), A(k)) \to H^0(S(l), A(l)) \to$$
$$\to H^1(S(g), A(g)) \to H^1(S(h), A(h)) \oplus H^1(S(k), A(k)) \to \dots$$

(Mayer-Vietoris sequence).

It should be remarked that all three statements are well-known (see [9]), but have now been exhibited as special cases of one and the same theorem.

As a further application we mention cohomology of groups relative to a family of subgroups. Take

$$E = \{e_i, g \mid i \in I, e_i \geq g, e_i, e_j \text{ incomparable, for all } i, j \in I\} ,$$

$S : E \to \underline{\underline{\mathrm{Gp}}}$ a semilattice of groups, and A a homogeneous S-module; moreover assume that $S(e_i \geq g)$ are monomorphisms. The same methods lead to an exact sequence

$$0 \to H^0(S(g), A(g)) \to \prod_{i \in I} H^0(S(e_i)) \to H^1_S(g, A) \to$$
$$\to H^1(S(g), A(g)) \to \prod_{i \in I} H^1(S(e_i), A(e_i)) \to \dots$$

which is the sequence established by Bautista Ramos ([2]) using different methods.

References

[1] Maurice Auslander, "Relative cohomology theory of groups and continuations of homomorphisms", Thesis, Columbia University, New York, 1954.

[2] Raymundo Bautista Ramos, "Note on the cohomology of a family of subgroups of a group G" (Spanish), *An. Inst. Mat. Univ. Nac. Autónoma México* **11** (1971), 33-42. MR**48**#4147.

[3] A.H. Clifford and G.B. Preston, *The algebraic theory of semigroups*, Vol. I (Mathematical Surveys, 7. Amer. Math. Soc., Providence, Rhode Island, 1961). MR**24**#A2527.

[4] A.H. Clifford and G.B. Preston, *The algebraic theory of semigroups*, Vol. II (Mathematical Surveys, 7. Amer. Math. Soc., Providence, Rhode Island, 1967). MR**36**#1558.

[5] Serge Lang, *Rapport sur la Cohomologie des Groupes* (Mathematics Lecture Notes. W.A. Benjamin, New York, Amsterdam, 1966). MR**35**#2948.

[6] H. Lausch, "Cohomology of inverse semigroups", *J. Algebra* **35** (1975), 273-303.

[7] H. Lausch, "Induced and coinduced modules for inverse semigroups", Algebra Paper **15** (1975), Monash University.

[8] H. Lausch, "Inverse semigroups and relative cohomology of groups", Algebra Paper **17** (1975), Monash University.

[9] Luis Ribes, "On a cohomology theory for pairs of groups", *Proc. Amer. Math. Soc.* **21** (1969), 230-234. MR**38**#4563.

[10] Luis Ribes, "Cohomological characterization of amalgamated products of groups", *J. Pure Appl. Algebra* 4 (1974), 309-317. Zbl M288#20067.

[11] Satoru Takasu, "Relative homology and relative cohomology theory of groups", *J. Fac. Sci. Univ. Tokyo Sect.* I 8 (1959), 75-110. MR22#1610.

Department of Mathematics,
Monash University,
Clayton,
Victoria, Australia.

PROC. MINICONF. THEORY OF GROUPS
CANBERRA 1975, 73-84. 20D15

DETERMINATION OF GROUPS OF PRIME-POWER ORDER

M.F. Newman

This talk is a report on work in progress both because the theory to be outlined has not been seriously tested in practice and because it has not been fully compared with related theories. The practical side of the problem is being undertaken by a student, Judith Ascione.

About a hundred years ago Cayley (1878) said:

"... only a little has been done towards the solution of the general problem ... . The general problem is to find all the groups of a given order n ; thus if $n = 2$, the only group is $1, \alpha$ $(\alpha^2 = 1)$; ..."

Much work has been done on this general problem since. The going has been getting progressively more difficult. All the work on classifying simple groups comes within the scope of it. My own present interest in the problem arises from having acquired a tool, a nilpotent quotient algorithm, which it seems might help in making further progress on the case of prime-power orders.

Of course the general problem can be solved in principle. For instance one could write down all Latin squares of a given order, check which can be regarded as tables for associative multiplications and then test for isomorphisms. What one wants are practical methods. I won't here go into all the methods that have been proposed or used. Let me just sketch, very briefly, results obtained for the case of prime-power orders. Cayley himself had already settled the case of prime order in 1854. Netto (1882) settled squares of primes; Young (1893) and Hölder (1893), independently, settled cubes and fourth powers. Bagnera (1898) essentially settled fifth powers (there are errors for 2^5 and 3^5 ; the former was pointed out by Miller (1899) and corrected by Bagnera (1899), the latter were pointed out by Bender

(1927) who himself was in error; the first complete list for 3^5 was that of James (1969); other lists have been calculated by de Séguier (1904) and, for $p \geq 5$, by Schreier (1926) in connection with his development of extension theory). In the 1930s P. Hall and Senior, at first independently and then jointly, worked on making a list of groups of order 2^n, $n \leq 6$. This finally appeared in book form in 1964 (M. Hall and Senior). The methods are described in chapters 3 and 4 of that book. The classification theory is described in the first of the papers (1940) associated with P. Hall's lectures to the Göttingen meeting of 1939. It is perhaps appropriate to quote from this paper:

> In order to "determine", that is, to construct, all groups of a given kind, it is necessary first to know something of the structure of the groups in question. The methods of the various authors who have dealt with this *construction problem* differ principally in the extent and nature of the *structure theorems* to which they make appeal. Obviously, the more one knows of the structure of a given class of groups, the easier (*ceteris paribus*) it should be to construct them. For instance, nearly everyone who has made determinations of prime-power groups has used the fact that in such a group there always exist subgroups whose index is a prime number, and that all such subgroups are self-conjugate. This fact ensures that every group G of order p^n (p a prime) can be obtained in at least one way by a simple extension of some group G' of order p^{n-1}:
>
> $$G = \{G', \xi\} \text{ , with } \xi^{-1}G'\xi = G' \text{ and } \xi^p \text{ in } G' .$$
>
> That G can in general be obtained by this method in many different ways is a very serious difficulty. A dictum of W. Magnus (1937) is here very much to the point. He writes: Die Hauptschwierigkeit besteht dabei nicht in einer Konstruktion aller Gruppen eines bestimmten Typs, sondern in der Angabe eines vollständigen Systems nicht isomorpher Gruppen aus den konstruierten Gruppen.

I might also quote his warning:

> To put it crudely, there is no apparent limit to the complication of a prime-power group. As we pass from the groups of order p^3 to those of order p^4, then to those of order p^5, and so on, at each step new structural phenomena make their appearance. And it seems unlikely that it will be possible to compass the overwhelming variety of prime-power groups within the bounds of a single finite system of formulae.

P. Hall also went some way into the case of groups of order 2^7. (I am indebted to him for making available to me summary tables for the completed part of that work.)

More recently these methods have been used by James in his thesis (1969) to the University of Sydney to make a list of all groups of order p^n, $n \le 6$. (I understand some of the detail was not completely accurate; and that James is at present reworking the list for publication.) As reported in a paper by James and Cannon (1969) this work was computer assisted to the extent that in some cases results for small primes were computed and used to point the way to the general result. Rodemich has worked on groups of order 2^7 and claims there are 2356 such groups. (No further details are available at present. I believe the use of a computer was involved.) Further progress can be made if restrictions are placed on the groups being considered; but that is another story.

What I will outline is a procedure which given a prime p and a positive integer d generates a list of descriptions of all d-generator finite p-groups.

For this discussion the descriptions will be given in terms of descriptions of normal subgroups of a free group F of rank d. Thus the procedure will be one for generating a list $\mathcal{R}$ of descriptions of normal subgroups of F such that

(1) for every R described in $\mathcal{R}$ the quotient group F/R is a d-generator finite p-group,

(2) for every d-generator finite p-group P there is exactly one N described in $\mathcal{R}$ such that F/N is isomorphic to P.

It will be convenient to introduce some terminology and notation.

Let H be a subgroup of a group G. The subgroup of G generated by all commutators $[h, g] = h^{-1}g^{-1}hg$ with h in H and g in G and all p-th powers h^p with h in H plays a vital role in what follows. It will be denoted $P(H, G)$. It may be described as a residual of H relative to G and could be called the exponent-p-central residual of H in G. Two basic properties are:

if $H \le K \le G$, then $P(H, G) \le P(K, G)$;

if θ is a homomorphism of G, then

$$P(H, G)\theta = P(H\theta, G\theta) .$$

This relative residual can be used to define an important descending chain of subgroups

$$G = P_0(G) \ge \dots \ge P_c(G) \ge \dots ,$$

the lower exponent-p-central chain, by

$$P_{c+1}(G) = P\big(P_c(G), G\big) .$$

The following simple properties of this chain are relevant to the later discussion.

(a) If $P_c(G) = P_{c+1}(G)$, then $P_{c+1}(G) = P_{c+2}(G)$.

(b) Each term $P_c(G)$ is fully invariant in G.

(c) If G is finitely generated, then $G/P_c(G)$ is a finite p-group.

(d) If P is a finite p-group, then there is an integer c such that $P_c(P) = E$ (the identity). If the lower exponent-p-central chain of a group G reaches E, then the least integer c such that $P_c(G) = E$ is the exponent-p-central class, or in this context simply class, of G.

(e) $G/P_c(G)$ has class at most c.

(f) If $G/P_c(G)$ has class c, then $G/P_{c-1}(G)$ has class $c - 1$.

(g) If G/N has class c, then $P_c(G) \leq N$ and $P_{c-1}(G) \not\leq N$.

Recall that a subgroup H of a group G is omissible if every subset S of G which with H generates G generates G by itself.

(h) If $P_c(G) = E$, then $P_1(G)$ is omissible in G.

(This is a special case of a more general result which is a natural analogue of Lemma 4 of McLain (1959). It can be proved in essentially the same way.)

The basic step in generating $\mathcal{R}$ is an algorithm N for calculating from the description $\mathcal{D}$ of a normal subgroup R of p-power index in F which is contained in $P_1(F)$ a set $N(\mathcal{D})$ of descriptions of normal subgroups of F contained in R and containing $P(R, F)$. The list $\mathcal{R}$ is the union of finite lists $\mathcal{R}_0, \mathcal{R}_1, \ldots, \mathcal{R}_\lambda, \ldots$ where $\mathcal{R}_0$ consists of a description $\mathcal{D}_1$ of $P_1(F)$;

$$\mathcal{R}_1 = \mathcal{R}_0 \cup N(\mathcal{D}_1)$$

$$\vdots$$

$$\mathcal{R}_{\lambda+1} = \mathcal{R}_\lambda \cup N(\mathcal{D}_\lambda) \quad \text{where } \mathcal{D}_\lambda \in \mathcal{R}_\lambda \backslash \{\mathcal{D}_1, \ldots, \mathcal{D}_{\lambda-1}\}$$

$$\vdots$$

Thus every description in $\mathcal{R}$ is a description of a normal subgroup of p-power index in F which is contained in $P_1(F)$. It follows that for every R described in $\mathcal{R}$ the quotient group F/R is a d-generator finite p-group. The other property of $\mathcal{R}$ cannot be established until more information has been given about N.

The algorithm N involves the use of the automorphism group, aut F/R, of the quotient group F/R under consideration. This automorphism group could be calculated from F/R. However in practice it is obtained along with R. So the algorithm is really one for calculating from a description of a normal subgroup R of F contained in $P_1(F)$ and a description of the automorphism group of F/R a set of descriptions each of a normal subgroup N of F contained in R and containing $P(R, F)$ and the corresponding automorphism group F/N . (The first two stages of the case $p = d = 2$ are outlined in an appendix.) From now on $P(R, F)$ will be denoted R^* . The description of R is in terms of a consistent power commutator presentation for F/R and that of aut F/R in terms of a set of generators given by their action on the set of generators of F/R . The first step of the algorithm calculates a description of R^* . This is done using the appropriate part of a nilpotent quotient algorithm. (Accounts of this have been given in the literature by Macdonald (1974) and Wamsley (1974) and in lectures in Canberra, Sydney and Aachen by me; there are implementations by Macdonald in Stirling, Bayes-Kautsky-Wamsley in Adelaide, W. Felsch in Aachen and Havas-Newman in Canberra.) It also yields a description of the elementary abelian group R/R^* . The second step of the algorithm calculates a description of $P_c(F/R^*)$ where c is the exponent-p-central class of F/R and a list of descriptions of the set U of proper supplements of $P_c(F/R^*)$ in R/R^* . Each automorphism of F/R extends to an automorphism of F/R^* . The third step of the algorithm calculates for each of the generating automorphisms β of F/R a description of an automorphism β^* of F/R^* which extends β and for each of these automorphisms β^* a description of the permutation it induces on U . (Since $R \leq P_1(F)$ and $[R, F]R^p = R^*$ this permutation depends only on β and not on the particular extension β^* .) The fourth step of the algorithm calculates a description of a set V consisting of one representative from each of the orbits in U under the action of the group generated by these permutations induced from aut F/R and for each N/R^* in V a description of a set of generators for the largest subgroup of aut F/R which induces permutations stabilizing N/R^* . These stabilizing automorphisms induce automorphisms in F/N which together with the automorphisms of F/N which induce the identity automorphism of F/R generate aut F/N . For each N/R^* in V one gets a description of N and of aut F/N in the required form. Observe that $NP_c(F) = R$ because N/R^* is a supplement of $P_c(F/R^*)$ in R/R^* ; hence, if N is described in R , then $NP_c(F)$ is also described in R .

Property (2) of R can now be established. Let P be a d-generator finite p-group. Induction on the exponent-p-central class of P will be used to establish that there is a normal subgroup N of F described in R with F/N isomorphic to P . If the class of P is 1 , then P is isomorphic to $F/P_1(F)$. If the class of P is $c + 1$ for some positive integer c , then $P/P_c(P)$ has class c and it can be assumed there is an R described in R such that F/R is isomorphic to $P/P_c(P)$. Let A be a d-element generating set for F and let θ be a homomorphism from F to $P/P_c(P)$ with kernel R. There is a theorem of Gaschütz (1956) which guarantees that P has a d-element generating set $\{b_a : a \in A\}$ such that $a\theta = b_a P_c(P)$. Let ψ be the homomorphism from F onto P which maps each a in A to the corresponding b_a and let M be the kernel of ψ, then F/M is isomorphic to P. Note that $R\psi = P_c(P)$, so $R/M = P_c(F/M)$. The following argument shows that M/R^* is in $\mathcal{U}$. Firstly

$$R^*\psi = P(R\psi, F\psi) = P\big(P_c(P), P\big) = E .$$

So $R^* \le M$. Clearly $MP_c(F) \le R$. Now $F/MP_c(F)$ has class at most c, so $R/M \le MP_c(F)/M$ and $R = MP_c(F)$. Finally $M \ne R$ because F/M has class $c + 1$ while F/R has class c. The construction of $\mathcal{V}$ guarantees there is an N described in $\mathcal{R}$ such that F/N is isomorphic to F/M and therefore to P. It remains to prove the uniqueness of N. Suppose L, N are described in $\mathcal{R}$ and F/L is isomorphic to F/N. To prove $L = N$ it suffices to find an automorphism of F/R whose extension to F/R^* maps L/R^* to N/R^*. This will be done by induction on the exponent-p-central class of F/L, F/N. If this class is 1, then $L = P_1(F) = N$. If the class is $c + 1$ for some positive integer c, then $LP_c(F)$, $NP_c(F)$ are described in $\mathcal{R}$ and $F/LP_c(F)$, $F/NP_c(F)$ are isomorphic and of class c; so it can be assumed that $LP_c(F) = NP_c(F) = R$. Let θ be an isomorphism from F/L to F/N. Put $K = P_{c+1}(F)$, then K is fully invariant in F and so there is an endomorphism ε of F/K such that $aK\varepsilon N = aL\theta$ for all a in the generating set A. This endomorphism ε can be shown to be an automorphism of F/K as follows. Since F/K is finite, it suffices to show ε is onto. Because θ is onto there is a subset $B^* = \{b_a^* : a \in A\}$ of F such that $b_a^*L\theta = aN$ for all a in A. Hence $b_a^*K\varepsilon N = b_a^*L\theta = aN$. Therefore $(B^*K/K)\varepsilon \cup N/K$ generates F/K. Since N/K is contained in $P_1(F)/K$ which is omissible in F/K, it follows that $(B^*K/K)\varepsilon$ generates F/K and therefore ε is onto as required. Moreover $(L/K)\varepsilon = N/K$: for if $w \in L$, then $wK\varepsilon N = wL\theta = N$ and so $(L/K)\varepsilon \le N/K$; but L/K, N/K have the same order and ε is an automorphism, so $(L/K)\varepsilon = N/K$. Also R/K admits ε for

$$(R/K)\varepsilon = \big[LP_c(F)/K\big]\varepsilon = NP_c(F)/K = R/K .$$

It follows that R^*/K admits ε. Therefore the automorphism of F/R induced by ε has the required property and the uniqueness of N is proved. (A similar argument can be found in Higman (1960), p. 26.) This completes the proof of property (2) of $\mathcal{R}$.

Some of the above results can be expressed in terms of automorphisms classes. A set $\mathcal{H}$ of subgroups of a group G is said to be an automorphism class of G if for each pair H, H^* of subgroups in $\mathcal{H}$ there is an automorphism δ of G such that $H\delta = H^*$. If P is a d-generator finite p-group of class at most c and if F is a d-generator free group, then the preceding discussion yields that the kernels of the homomorphisms from $F/\mathcal{P}_c(F)$ onto P form an automorphism class of $F/\mathcal{P}_c(F)$. It is not difficult to see that this remains true when the rank of F is greater than d. The above proofs show that the required automorphisms are all induced from endomorphisms of F. They are not always induced from automorphisms of F for Neumann (1956) has described a 2-generator 2-group P for which the kernels of homomorphisms from a free group of rank 2 onto P form at least two automorphism classes in F.

It is not clear to me how this method would compare with that described by P. Hall and M. Hall and Senior as a basis for computer calculations. That method is not described in an algorithmic form and I have not, as yet, written it in such a form. In that method groups are classified into families, using isoclinism; each family is divided into branches; and then all the groups in one branch of a family are calculated. All the groups in a family have the same central quotient group. Given a finite p-group P there are only finitely many families with P as central quotient group. If P is isomorphic to F/R, the outer automorphisms of P are represented in a natural way by automorphisms of $R \cap F'/[R, F]$ (the Schur multiplier of P). The subgroups N of F such that $[R, F] \leq N \leq R \cap F'$ and such that R/F is the centre of F/N are permuted among themselves in this representation. The families are in one-one correspondence with the transitivity sets of such subgroups. (The last three sentences are essentially Theorem 2.2 on p. 12 of Hall and Senior (1964).)

Sims is reported (Neubüser (1970), §2.1.2) to have described a "procedure by which in principle each group of prime-power order would be obtained just once", to have "written a programme along these lines which determined the two-generator groups of order 32 in a very short time", and to have expressed the hope that most groups of order 128 would be determined by this procedure in a reasonable time. (In June 1975 Sims said (privately) that he had not continued with this project.)

Another method has been proposed by Leedham-Green. (This is not yet written down in detail and I am indebted to Dr Leedham-Green for sending me an outline of his method.) A special case of this method has been programmed for a computer by a student, Maung, of Leedham-Green to calculate information about 5-groups of maximal class. The method is quite close to the one described above. It differs only in the choice of the set $\mathcal{U}$ above. Given P it obtains one copy of each group Q which has no direct factor which is cyclic of order p and is such that $Q/\mathcal{P}^*(Q)$ is isomorphic to P - here $\mathcal{P}^*(Q)$ is the subgroup generated by all central elements of order p in Q.

APPENDIX

An outline of the beginning of the procedure for $p = d = 2$.

F is free on $\{a_1, a_2\}$.

The description $\mathcal{D}_1$ of $P_1(F)$ (or simply P_1) and aut F/P_1 is:

$$a_1^2 = e \text{ , } a_2^2 = e$$

$$[a_2, a_1] = e$$

(that is P_1 is the normal closure in F of a_1^2, a_2^2, $[a_2, a_1]$);

	a_1P_1	a_2P_1
β_{11}	a_1P_1	$a_1a_2P_1$
β_{12}	$a_1a_2P_1$	a_2P_1 .

Step 1. Calculation of $P(P_1, F) = P_2(F) = P_2$ and P_1/P_2 .

Define

$$a_3 = [a_2, a_1] \text{ , } a_4 = a_1^2 \text{ , } a_5 = a_2^2 \text{ ;}$$

P_2 is described by

$$a_1^2 = a_4 \text{ , } a_2^2 = a_5 \text{ , } a_3^2 = a_4^2 = a_5^2 = e$$

$$[a_2, a_1] = a_3 \text{ ,}$$

$$[a_3, a_1] = [a_3, a_2] = [a_4, a_1] = [a_4, a_2] = [a_5, a_1] = [a_5, a_2] = e \text{ ;}$$

P_1/P_2 is generated by

$$a_3P_2, a_4P_2, a_5P_2 \text{ .}$$

Step 2. Calculation of $P_1(F/P_2)$ and proper supplements.

$P_1(F/P_2)$ is generated by

$$a_3P_2, a_4P_2, a_5P_2 \text{ ;}$$

the proper supplements are:

$$U_1 = \langle a_4P_2, a_5P_2\rangle, \quad U_2 = \langle a_3a_4P_2, a_5P_2\rangle, \quad U_3 = \langle a_3P_2, a_5P_2\rangle,$$
$$U_4 = \langle a_3P_2, a_4a_5P_2\rangle, \quad U_5 = \langle a_3a_5P_2, a_4a_5P_2\rangle, \quad U_6 = \langle a_4P_2, a_3a_5P_2\rangle,$$
$$U_7 = \langle a_3P_2, a_4P_2\rangle, \quad U_8 = \langle a_5P_2\rangle, \quad U_9 = \langle a_4a_5P_2\rangle,$$
$$U_{10} = \langle a_3a_4a_5P_2\rangle, \quad U_{11} = \langle a_3a_5P_2\rangle, \quad U_{12} = \langle a_4P_2\rangle,$$
$$U_{13} = \langle a_3a_4P_2\rangle, \quad U_{14} = \langle a_3P_2\rangle, \quad U_{15} = \langle P_2\rangle .$$

Step 3. Calculation of extended automorphisms and permutations.

	a_1P_2	a_2P_2
β_{11}^*	a_1P_2	$a_1a_2P_2$
β_{12}^*	$a_1a_2P_2$	a_2P_2 ;

$$(U_1U_6)(U_3U_4)(U_8U_{10})(U_9U_{11})$$
$$(U_1U_2)(U_4U_7)(U_9U_{13})(U_{10}U_{12}) .$$

Step 4. Calculation of orbit representatives and stabilizers.

Orbit representatives are

$$U_1, U_3, U_5, U_8, U_9, U_{14}, U_{15} ;$$

$$\mathrm{stab}(U_1) = \langle \beta_{11}\beta_{12}\beta_{11}\rangle, \quad \mathrm{stab}(U_3) = \langle \beta_{12}\rangle, \quad \mathrm{stab}(U_5) = \langle \beta_{11}, \beta_{12}\rangle,$$
$$\mathrm{stab}(U_8) = \langle \beta_{12}\rangle, \quad \mathrm{stab}(U_9) = \langle \beta_{11}\beta_{12}\beta_{11}\rangle, \quad \mathrm{stab}(U_{14}) = \langle \beta_{11}, \beta_{12}\rangle,$$
$$\mathrm{stab}(U_{15}) = \langle \beta_{11}, \beta_{12}\rangle .$$

From this one gets descriptions for the seven two-generator 2-groups G with $P_1(G) > P_2(G) = E$ and their automorphism groups. The first of these, $\mathcal{D}_2$, follows; it describes the dihedral group of order 8 .

$$a_3 = [a_2, a_1] ,$$
$$a_1^2 = a_2^2 = a_3^2 = e ,$$
$$[a_2, a_1] = a_3 , \quad [a_3, a_1] = [a_3, a_2] = e ;$$

	a_1R	a_2R
β_{21}	a_1a_3R	a_2R
β_{22}	a_1R	a_2a_3R
β_{23}	a_2R	a_1R

(the first two induce the identity on F/P_1, the last induces $\beta_{11}\beta_{12}\beta_{11}$ on F/P_1).

Now N is applied to $\mathcal{D}_2$.

Step 1. R^* and R/R^* .

$$a_4 = [a_3, a_1] , \quad a_5 = a_1^2 , \quad a_6 = a_2^2 ,$$

$$a_1^2 = a_5 , \quad a_2^2 = a_6 , \quad a_3^2 = a_4 , \quad a_4^2 = a_5^2 = a_6^2 = e$$

$$[a_2, a_1] = a_3 , \quad [a_3, a_1] = [a_3, a_2] = a_4 , \quad [a_j, a_i] = e \text{ otherwise;}$$

$$a_4R^*, a_5R^*, a_6R^* .$$

Step 2. $P_2(F/R^*)$ and U .

$$a_4R^* ;$$

$$U_1 = \langle a_5R^*, a_6R^* \rangle , \qquad U_2 = \langle a_4a_5R^*, a_6R^* \rangle ,$$

$$U_3 = \langle a_4a_6R^*, a_5a_6R^* \rangle , \quad U_4 = \langle a_5R^*, a_4a_6R^* \rangle .$$

Step 3. Permutations.

$$(U_2U_4) .$$

Step 4. Orbit representatives and stabilizers.

$$U_1, U_2, U_3 ;$$

$$\text{stab}(U_1) = \langle \beta_{21}, \beta_{22}, \beta_{23} \rangle , \quad \text{stab}(U_2) = \langle \beta_{21}, \beta_{22} \rangle , \quad \text{stab}(U_3) = \langle \beta_{21}, \beta_{22}, \beta_{23} \rangle .$$

From this one gets descriptions for the three two-generator groups G with $P_2(G) > P_3(G) = E$ and $G/P_2(G)$ dihedral of order 8 and their automorphism groups. The groups are, respectively, dihedral, semi-dihedral and quaternion of order 16 .

Now the procedure would continue with the second of the groups created at the first stage. And so on till space, time or patience runs out. It is perhaps worth observing that the third of the groups created at the first stage (which is the quaternion group) has no extensions (in this sense).

References

G. Bagnera (1898), "La composizione dei Gruppi finiti il cui grado è la quinta potenza di un numero primo", *Ann. Mat. Pura Appl.* (3) 1, 137-338. FM**29**,112.

G. Bagnera (1899), "Sopra i gruppi astratti di grado 32 ", *Ann. Mat. Pura Appl.* (3) **2**, 263-275. FM**30**,133.

H.A. Bender (1927), "A determination of the groups of order p^5 ", *Ann. of Math.* (2) **29**, 61-72. FM**53**,105.

A. Cayley (1854), "On the theory of groups, as depending on the symbolic equation $\theta^n = 1$ ", *Philos. Mag.* 7, 40-47. See also: *The Collected Mathematical Papers of Arthur Cayley*, Vol. II, #125, 123-130 (Cambridge University Press, Cambridge, 1889).

A. Cayley (1878), "Desiderata and suggestions. No. 1. The theory of groups", *Amer. J. Math.* 1, 50-52. See also: *The Collected Mathematical Papers of Arthur Cayley*, Vol. X, #694, 401-403 (Cambridge University Press, Cambridge, 1896).

Wolfgang Gaschütz (1956), "Zu einem von B.H. und H. Neumann gestellten Problem", *Math. Nachr.* 14, 249-252 (1955). MR18,790.

Marshall Hall, Jr. and James K. Senior (1964), *The Groups of Order* 2^n $(n \leq 6)$ (Macmillan, New York; Collier-Macmillan, London). MR29#5889.

P. Hall (1940), "The classification of prime-power groups", *J. reine angew. Math.* 182, 130-141. MR2,211.

Graham Higman (1960), "Enumerating p-groups. I. Inequalities", *Proc. London Math. Soc.* (3) 10, 24-30. MR22#4779.

Otto Hölder (1893), "Die Gruppen der Ordnungen p^3, pq^2, pqr, p^4 ", *Math. Ann.* 43, 301-412. FM25,201.

Rodney K. James (1969), "The groups of order p^6 $(p \geqq 3)$ ", PhD thesis, University of Sydney, Sydney.

Rodney James and John Cannon (1969), "Computation of isomorphism classes of p-groups", *Math. Comp.* 23, 135-140. MR39#313.

I.D. Macdonald (1974), "A computer application to finite p-groups", *J. Austral. Math. Soc.* 17, 102-112. ZblM277#20024.

D.H. McLain (1959), "Finiteness conditions in locally soluble groups", *J. London Math. Soc.* 34, 101-107. MR21#2003.

Wilhelm Magnus (1937), "Neuere Ergebnisse über auflösbare Gruppen", *Jber. Deutsche. Math.-Verein.* 47, 69-78. ZblM16,202.

G.A. Miller (1899), "Report on recent progress in the theory of the groups of a finite order", *Bull. Amer. Math. Soc.* 5, 227-249 [p. 246]. See also: *The Collected Works of George Abram Miller*, Volume I, #48, 326-344 [p. 342] (University of Illinois Urbana, Illinois, 1935).

E. Netto (1882), *Substitutionentheorie und ihre Anwendungen auf die Algebra* (Teubner, Leipzig). FM14,90.

J. Neubüser (1970), "Investigations of groups on computers", *Computational Problems in Abstract Algebra* (Proc. Conf. Oxford, 1967, 1-19. Pergamon Press, Oxford, London, Edinburgh, New York, Toronto, Sydney, Paris, Braunschweig). MR41#5480.

B.H. Neumann (1956), "On a question of Gaschütz", *Arch. Math.* 7, 87-90. MR18,11.

Otto Schreier (1926), "Über die Erweiterung von Gruppen. II", *Abh. Math. Sem. Univ. Hamburg* 4, 321-346. FM52,113.

J.-A. Séguier (1904), *Théorie des groupes finis. Éléments de la théorie des groupes abstraits* (Gauthier-Villars, Paris).

J.W. Wamsley (1974), "Computation in nilpotent groups (theory)", *Proc. Second Internat. Conf. Theory of Groups*, Canberra 1973, 691-700 (Lecture Notes in Mathematics, 372. Springer-Verlag, Berlin, Heidelberg, New York). MR50#7350.

J.W.A. Young (1893), "On the determination of groups whose order is a power of a prime", *Amer. J. Math.* 15, 124-178. FM25,201.

Department of Mathematics,
Institute of Advanced Studies,
Australian National University,
Canberra, ACT, Australia.

PROC. MINICONF. THEORY OF GROUPS
CANBERRA 1975, 85-111.

05B25
(20H15)

ON PARTIALLY TRANSITIVE PROJECTIVE PLANES OF CERTAIN HUGHES TYPES

Cheryl E. Praeger and Alan Rahilly

1. Introduction

Consider a finite projective plane π with a collineation group G acting on it which fixes elementwise a substructure π' of π, and which acts regularly on the set of points incident with no line of π', and on the set of lines incident with no point of π'. We shall refer to the points and lines in these sets as *ordinary points* and *ordinary lines*, respectively. The elements (points and lines) of π' will be referred to as *fixed*, elements incident with a unique element of π' will be referred to as *tangent*, and a plane of the above type shall be called *partially transitive relative to* π'.

The structure π' must be a subplane of π (degenerate or non-degenerate) and Hughes [5] has exhaustively listed the various types that might occur. In this paper we wish to consider three of the types listed by Hughes, namely types $(4, m)$, $(5, m)$ and $(6, m)$.

1. Type $(4, m)$. π' consists of m points $(m \geq 3)$ Q_i, $i = 1, 2, \ldots, m$, on a line K_0, and a point Q_0 not on K_0, and the $m+1$ lines K_0, $K_i = Q_0Q_i$, $i = 1, \ldots, m$.

2. Type $(5, m)$. π' consists of $m+1$ points $(m \geq 2)$ Q_i, $i = 0, 1, \ldots, m$, on a line K_0, and $m+1$ lines K_i, $i = 0, 1, \ldots, m$ each through Q_0.

3. Type $(6, m)$. π' consists of m^2+m+1 points Q_i and lines K_i $(i = 0, 1, \ldots, m^2+m)$ which constitute a subplane of π of order m.

In his paper [5] Hughes shows that an infinite family of finite planes of type

$(5, m)$ exist (namely, the finite Hall planes). In a second paper [6], published shortly after, he established that the same class of planes is an infinite family of type $(4, m)$. In a footnote of [5] (pp. 652-653) and also later in the paper (p. 673) he claims not to know of any planes of type $(6, m)$. Recently Lorimer (see [15]) has constructed a translation plane of order 16 which Johnson and Ostrom have shown to be of type $(6, 2)$ (see [9]). This plane was constructed independently by Rahilly ([15], [17] and [18]). Furthermore, Rahilly and Searby [19] have shown that, if π is a partially transitive plane of Hughes type $(6, p)$, where p is a prime, then $p \leq 3$. It is the aim of this paper firstly, to characterize the translation planes of types (x, m), $x = 4, 5, 6$ and, secondly, to investigate in some detail the action of the group G on an arbitrary plane π of type (x, m), $x = 4, 5, 6$. A point of departure for our work is provided by the work of Hughes [5] which we shall summarize in the next section.

2. Preliminaries

Let π be a partially transitive plane of Hughes type (x, m), $x = 4, 5, 6$, of order n. Then $n = (m-1)^2$, m^2 or m^4, as $x = 4, 5, 6$. Further, the action of G on π is as follows:

(a) If π is of type $(4, m)$, then the stabilizer in G of a point on K_i, $i = 1, \ldots, m$, fixes pointwise a Baer subplane of π containing π' and is regular on the set of points on any line of the subplane which are not in the subplane. Also, each such stabilizer has exactly m conjugates each of which fixes pointwise a Baer subplane containing π' and which acts regularly on the set of points on a line of the Baer subplane not in the subplane. Further, each pair of the Baer subplanes fixed pointwise by such a stabilizer intersect precisely in π', and the situation is that each line K_i, $i = 1, \ldots, m$, is incident with $m - 2$ of the points of each Baer subplane distinct from Q_0 and Q_i. We see that the points of π in the Baer subplanes are precisely those on the lines K_i, $i = 1, \ldots, m$.

We shall denote the m Baer subplanes, whose points constitute the lines K_i, by π_j, $j = 1, \ldots, m$, and the set $\{\pi_j \mid j = 1, \ldots, m\}$ by Ω.

(b) If π is of type $(5, m)$, then the stabilizer in G of a point on K_i, $i = 1, \ldots, m$, fixes pointwise a Baer subplane containing π' and acts regularly in the same way as mentioned in (a). Also, each such stabilizer has exactly m conjugates each of which fixes pointwise a Baer subplane containing π' and which acts regularly in the usual way. Further, each pair of such Baer subplanes intersect precisely in π', and each line K_i, $i = 1, \ldots, m$, is incident with exactly m of the points of each of these Baer subplanes distinct from Q_0. We see that the points of π in the Baer subplanes are precisely those on the lines K_i, $i = 1, \ldots, m$.

We shall denote the m Baer subplanes constituting the lines K_i by π_j, $j = 1, \ldots, m$, and the set $\{\pi_j \mid j = 1, \ldots, m\}$ by Ω.

(c) If π is of type $(6, m)$, then the stabilizer in G of a point of K_i, $i = 0, \ldots, m^2+m$, fixes pointwise a Baer subplane containing π' and acts regularly in the usual manner. Also, each such stabilizer has exactly $m^2 + m + 1$ conjugates each of which fixes pointwise a Baer subplane containing π', and which acts regularly on the set of points on any line of the subplane not in the subplane. Further, each pair of these Baer subplanes intersect precisely in π' and each line K_i, $i = 0, \ldots, m^2+m$, is incident with exactly $m(m-1)$ of the points of each of these Baer subplanes not in π'. We see that the points of π in the Baer subplanes are precisely those on the lines K_i, $i = 0, \ldots, m^2+m$.

We shall denote the $m^2 + m + 1$ Baer subplanes constituting the lines K_i by π_j, $j = 0, \ldots, m^2+m$, and the set $\left\{\pi_j \mid j = 0, \ldots, m^2+m\right\}$ by Ω.

In each of the three cases the group G acts doubly transitively on Ω. We shall denote the subgroup of G fixing π_j pointwise (which we note is the stabilizer of any point of $\pi_j \backslash \pi'$) by R_j and the setwise stabilizer of π_j by G_j. We let K be the kernel of the action of G on Ω, that is $K = \bigcap_{i \in \Omega} G_i$, and then denote the constituent of a subgroup H of G on the set Ω by $H^\Omega = HK/K$. The set of points of Ω fixed by H is denoted by $\mathrm{fix}_\Omega H$. Most of the notation used to describe the permutation actions is standard and the reader is referred to Wielandt's book [22]. The table on p.101 contains information concerning the orders of various groups. The orders of π and Ω are included for the sake of completeness.

We note in passing that the R_i are a T.I. set; that is, if $i \neq j$ then $R_i \cap R_j = 1$ since they fix distinct Baer subplanes pointwise. Notice also that a group R_i cannot occur in a desarguesian plane unless $|\pi| = 4$. The desarguesian plane of order four is not of type $(6, m)$. It is, however, of types $(4, 3)$ and $(5, 2)$.

x	4	5	6
$\lvert G\rvert$	$m(m-1)(m-2)^2$	$m^3(m-1)$	$m^3(m^2-1)(m-1)(m^2+m+1)$
$\lvert G_i\rvert$	$(m-1)(m-2)^2$	$m^2(m-1)$	$m^3(m^2-1)(m-1)$
$\lvert R_i\rvert$	$(m-1)(m-2)$	$m(m-1)$	$m^2(m^2-1)$
$\lvert R_i\cap G_j\rvert$	$m-2$	m	$m(m-1)$
$\lvert \pi\rvert$	$(m-1)^2$	m^2	m^4
$\lvert \Omega\rvert$	m	m	m^2+m+1

3. Finite generalized Hall planes

A projective plane π is said to be a *generalized Hall plane with respect to* l_∞, π_0 if

(i) π is a translation plane with axis l_∞ ,

(ii) π_0 is a Baer subplane containing l_∞ ,

(iii) there is a collineation group $G(\pi_0)$ of π which fixes π_0 pointwise and which is regular on the point set $l_\infty \backslash \underline{\underline{M}}$, where $\underline{\underline{M}} = l_\infty \cap \pi_0$.

In a generalized Hall plane the subplane π_0 is desarguesian and the plane may be coordinatized over a quadrangle $0, I, X, Y$ in π_0 , such that $XY = l_\infty$, by a quasifield F with the properties

(a) F is a right vector space of dimension two over the subfield F_0 coordinatizing π_0 ,

(b) $(z\alpha+\beta)z = z\{f(\alpha)+h(\beta)\} + g(\alpha) + k(\beta)$ for all $\alpha, \beta \in F_0$, $z \in F\backslash F_0$, where f, g, h and k are endomorphisms of $(F_0, +)$ satisfying

   (i) $h(1) = 1$, $k(1) = 0$,

   (ii) $h \in \mathrm{Aut}(F_0, +)$,

   (iii) $M_\lambda = g + (k-m_\lambda)h^{-1}(m_\lambda-f) \in \mathrm{Aut}(F_0, +)$ for all $\lambda \in F_0$, where m_λ denotes the endomorphism $m_\lambda : F_0 \to F_0 : x \to \lambda x$ of $(F_0, +)$.

Such a quasifield F is called a *generalised Hall system* and the endomorphisms f, g, h, k the *defining functions* of F . Note that a generalized Hall system is a

Hall system (Hall [7], p. 364) if and only if $f = m_r$, $g = m_s$, $h = 1$, $k = 0$, where $x^2 - rx - s$ is irreducible over F_0 .

Finite generalized Hall planes are derivable planes, the set $\underline{\underline{M}}$ being a derivation set. The derived planes are semifield planes (of type 1 - $3a$ of Johnson's classification of semi-translation planes - see [8]) which are desarguesian planes if and only if the generalized Hall plane is a Hall plane.

An important result on the action of the full collineation group of a finite generalized Hall plane π is that, if $|\pi| > 16$, then $\underline{\underline{M}}$ is fixed by all collineations. This is Theorem 3.3.9 of [15] and also Theorem 1 of [16].

There are precisely three generalized Hall planes of order 16 . We shall need some information about the orbit structure of the full collineation group on l_∞ for these planes.

1. **The Hall plane.** The endomorphisms $f = m_t$, $g = h = 1$, $k = 0$ of GF(4) , where $t^2 = t + 1$, are the defining functions of a Hall system of order 16 . The orbit structure of the full collineation group of the plane ρ coordinatized by this quasifield is $\underline{\underline{M}}$, $l_\infty \backslash \underline{\underline{M}}$ and so the orbit lengths on l_∞ are 5 and 12 .

2. **The Lorimer plane.** In his paper [11] Lorimer has constructed a remarkable translation plane of order 16 which we shall denote by σ . Independently the same plane was constructed by Rahilly ([17]). The plane σ is a generalized Hall plane which is coordinatized by a generalized Hall system with defining functions $f = g = \phi$, $h = 1$, $k = 1 + \phi$, where ϕ is the non-trivial automorphism of GF(4) . Johnson and Ostrom [9] have shown that this plane is of Hughes type (6, 2) . A detailed discussion of this plane appears in [18]. The orbit decomposition of the full collineation group on l_∞ is $l_\infty \cap \pi'$, $l_\infty \backslash (l_\infty \cap \pi')$, where π' is a Fano subplane of σ relative to which σ is partially transitive of Hughes type (6, 2) . Thus the orbit lengths on l_∞ are 3, 14 .

3. **A third plane.** There is one other generalized Hall plane of order 16 . It is coordinatized by a generalized Hall system with defining functions $f = 0$, $g = m_t\phi$, $h = \phi$, $k = 0$, where ϕ is the non-trivial automorphism of GF(4) and $t^2 = t + 1$. The middle and right nucleus of this quasifield are both isomorphic to $F_0 = \mathrm{GF}(4)$. It follows that the plane T coordinatized by it is (X, OY, π_0) - and (Y, OX, π_0) - transitive. The distributor of the generalized Hall system just defined is simply $\{0\}$ and so there are no non-trivial (Y, OY)-elations. In fact, there are no non-trivial affine axis elations at all. To see this suppose first of all that there is a non-trivial (P, OP)-elation e_1 , where $P \in \underline{\underline{M}} \backslash \{X, Y\}$. Then

e_1 fixes $\underline{\underline{M}}$ ([17], Lemma 6) and must swap X and Y. Without loss of generality we can take e_1 to be a $\big((1), O(1)\big)$-elation. But then Lemma 7 of [17] tells us that there is $\mu \in F_0\backslash\{0\}$ such that

$$(xz)(z+\mu) = x$$

for all $x \in F_0$. But $xz = z\phi(x)$ and so we have

$$\big((z+\mu)\phi(x)+\mu\phi(x)\big)(z+\mu) = x ;$$

that is,

$$(z+\mu)\phi\big(\mu\phi(x)\big) + tx = x .$$

Since this must be true for all $x \in F_0$ we have that $\mu = 0$, a contradiction.

Next, suppose there is a non-trivial (P, OP)-elation e_2 where $P \notin \underline{\underline{M}}$. Now e_2 must swap X and Y and cannot fix $\underline{\underline{M}}$. It follows that there must be a non-trivial (Q, OQ)-elation with $Q \in \underline{\underline{M}}\backslash\{X, Y\}$. But we have shown that this does not occur.

The possible orbit length decompositions on l_∞ for the full collineation group of T are

(i) 1, 1, 3, 12 ,

(ii) 1, 1, 15 ,

(iii) 2, 3, 12 ,

(iv) 2, 15 .

For X and Y are either both fixed, constitute an orbit or are part of an orbit of length 14. Otherwise we would have non-trivial affine axis elations. But if the orbit length decomposition was 3, 14, then there would be seven Baer subplanes relative to which T would be a generalized Hall plane, each containing the orbit of length 3 and 0. However any two of these Baer subplanes would have to overlap on a further point Q on OP, where P belongs to the orbit of length 3. Clearly, the intersection of any two of the Baer subplanes is a Fano subplane. But then T would be tangentially transitive relative to this Fano subplane and by the main result of Johnson and Ostrom [17] we would have T isomorphic to σ. This eliminates 3, 14 as a possible orbit length decomposition.

4. Translation planes of Hughes types (x, m), $x = 4, 5, 6$

The main purpose of this section is to establish the following theorem:

THEOREM 1. *(i) A finite translation plane* π *with axis* l_∞*, of order greater than nine, and not equal to* T*, is of Hughes type* $(4, m)$ *if and only if* π

is a Hall plane and $K_0 = l_\infty$.

(ii) A finite translation plane π *with axis* l_∞ *, of order greater than nine, is of Hughes type* $(5, m)$ *if and only if* π *is a generalized Hall plane and* $K_0 = l_\infty$.

(iii) A finite translation plane π *is of Hughes type* $(6, m)$ *if and only if* π *is the generalized Hall plane* σ *of order* 16 .

Proof. Hughes [6] has shown that the Hall planes are of type $(4, m)$ with $K_0 = l_\infty$. (See §5 of this paper for some details of the partially transitive group.)

Suppose π is a translation plane of Hughes type $(4, m)$ with translation axis l_∞ . If l_∞ is not fixed by G then π is desarguesian and so of order four. Since we have supposed $|\pi| > 9$ we have that l_∞ is fixed by G . Now this means either $l_\infty = K_i$ for some $i = 1, \ldots, m$ or $l_\infty = K_0$. But if $l_\infty = K_i$ for some $i = 1, \ldots, m$, then there are m Baer subplanes π_j containing l_∞ such that $(\pi_j \cap \pi_k) \cap l_\infty = \{Q_0, Q_i\}$, $j \neq k$, and with respect to which π is a generalized Hall plane. But since $\pi_j \cap l_\infty$ is not fixed by all collineations we must have $|\pi| = 16$, and since we have explicitly excluded the plane T we have $\pi = \rho$ or $\pi = \sigma$. However, the respective orbit length decompositions of these planes are 5, 12 and 3, 14 . But G is transitive on the set of points $l_\infty \backslash \{Q_0, Q_i\}$, and so we have a contradiction.

If $l_\infty = K_0$, then there are m Baer subplanes π_j containing l_∞ , such that $\pi_j \cap l_\infty = \{Q_i \mid i = 1, \ldots, m\}$ and with respect to which π is a generalized Hall plane. Now we can derive π using the derivation set $\underline{M} = \pi_j \cap l_\infty$ to obtain a semifield plane $\overline{\pi}$. The points on the m subplanes π_j are the points on the m lines $Q_0\overline{Q}_i$ of $\overline{\pi}$, where $\overline{Q}_i$ is the point of $\overline{\pi}$ replacing Q_i after derivation. But G permutes the subplanes π_j transitively, and so the group $\overline{G}$ induced by G on $\overline{\pi}$ is transitive on the $\overline{Q}_i$. Since $\overline{\pi}$ is a semifield plane with respect to one of the $\overline{Q}_i$ it follows that $\overline{\pi}$ is desarguesian. However, π may be derived from $\overline{\pi}$ by reversing the previous derivation, and so must be a Hall plane.

(ii) Let π be a finite generalized Hall plane with respect to l_∞ and π_0 of order greater than four. Suppose π is coordinatized over O, I, X, Y by a generalized Hall system F , π_0 being coordinatized by the subfield F_0 of F .

The group $T(Y)$ of translations with centre Y on l_∞ consists of the collineations

$$t_a : \pi \to \pi : \begin{cases} (x, y) \to (x, y+a) , \\ (m) \to (m) , \\ Y \to Y . \end{cases}$$

Also, the group $G(\pi_0)$ which fixes π_0 pointwise consists of collineations

$$\phi_{r,s} : \pi \to \pi : \begin{cases} (x, y) \to \left(\phi_{r,s}(x), \phi_{r,s}(y)\right) , \\ (m) \to \left(\phi_{r,s}(m)\right) , \\ Y \to Y , \end{cases}$$

where $\phi_{r,s}(z\alpha+\beta) = zr\alpha + s\alpha + \beta$ and $r, s \in F_0$.

We shall show that the group G generated by $T(Y)$ and $G(\pi_0)$ is partially transitive of Hughes type $(5, m)$ on π .

Firstly, $T(Y)$ is normal in G since all elements of G fix Y and l_∞ . Also $T(Y) \cap G(\pi_0) = 1$ and so $G = T(Y)G(\pi_0)$, the semi-direct product of T and $G(\pi_0)$. Notice that $|G| = m^2(m^2-m)$, where $m = |\pi_0|$. Next, it is easy to see that $\underline{\underline{M}} = l_\infty \cap \pi_0$ is the set of fixed points of G , and the lines of π_0 incident with Y are the fixed lines of G . Finally, it is easy to see that an arbitrary element $t_a\phi_{r,s}$ of G cannot fix a point on a line of $\pi\backslash\pi_0$ incident with Y . So G is semi-regular on the $m^2(m^2-m)$ points of this type. But $|G| = m^2(m^2-m)$ and therefore G is regular on the set of ordinary points relative to its fixed substructure. A further single argument shows that G is regular on the ordinary lines relative to its fixed substructure. (In [5] Hughes pointed out that this group is partially transitive of type $(5, m)$ on the Hall planes.)

Suppose π is a finite translation plane with axis l_∞ , of order greater than four, and of Hughes type $(5, m)$. As in part (a) we see that G fixes l_∞ and so $l_\infty = K_i$, for some $i = 1, \ldots, m$ or $l_\infty = K_0$. Now, if $l_\infty = K_i$, for some $i = 1, \ldots, m$, then there are m Baer subplanes π_j containing l_∞ , such that $(\pi_j \cap \pi_k) \cap l_\infty = \{Q_0\}$, $j \neq k$, and with respect to which π is a generalized Hall plane. But G does not fix $l_\infty \cap \pi_j$ and so, as before, $|\pi| = 16$. However, G is transitive on $l_\infty\backslash\{Q_0\}$ and this conflicts with the known orbit length decompositions on l_∞ of ρ and σ , and also with the possible orbit length decompositions on l_∞ for T . It follows that $l_\infty = K_0$ and we have m Baer subplanes π_j containing $\{Q_i \mid i = 0, \ldots, m\}$ with respect to which π is a generalized Hall plane.

(iii) The plane σ is a translation plane of Hughes type $(6, 2)$ - see

Johnson and Ostrom [9] or Rahilly [18], Theorem 7.

The converse part is Theorem 9 of Rahilly and Searby [19].

REMARKS. (a) The desarguesian plane of order four is a Hall plane (coordinatized by a Hall system with defining functions $f = g = h = 1$ and $k = 0$ on GF(2)) and so is of Hughes type (4, 3) and (5, 2) . Also, the Hall plane of order 9 is of types (4, 4) and (5, 3) .

(b) To the knowledge of the authors the full collineation group of the plane T is yet to be determined. It is not clear whether T is of Hughes type (4, 5) with $K_0 \neq l_\infty$ or not of Hughes type (4, 5) at all.

5. Partially transitive groups

In this section we consider some examples of "partially transitive" groups on finite projective planes.

1. Type (4, m). Consider the desarguesian plane $\overline{\pi}$ of order p^{2t} and suppose $\overline{\pi}_0$ is a Baer subplane of $\overline{\pi}$. If we coordinatize $\overline{\pi}$ over a quadrangle in $\overline{\pi}_0$ by the field $F = \mathrm{GF}(p^{2t})$, then the points of $\overline{\pi}_0$ will receive coordinates from a subfield $F_0 = \mathrm{GF}(p^t)$. There is a group of collineations G fixing $\overline{\pi}_0$ consisting of collineations (affine part only indicated)

$$(x, y) \to (xd_1+yd_2, xe_1+ye_2)$$

where $d_1, d_2, e_1, e_2 \in F_0$. Applying the process of derivation of planes and using the derivation set $l_\infty \cap \overline{\pi}_0$ we obtain the Hall plane π of order p^{2t} , and we see that $\overline{G}$ induces a group G of collineations in π consisting of collineations (affine part only again)

$$(x, y) \to (\psi(x), \psi(y))$$

where $\psi(z\alpha+\beta) = z(d_1\alpha+d_2\beta) + e_1\alpha + e_2\beta$, $z \notin F_0$, $\alpha, \beta \in F_0$. This group G is the one Hughes showed to be a partially transitive group of type $(4, p^t+1)$ in his paper [6].

Let us consider the action of this group on the Hall planes in more detail.

The lines K_i , $1 \leq i \leq p^t+1$, in π are those given by $y = x\xi$, $\xi \in F_0$, and also the line $x = 0$. The tangent points are (α, β) and $(w\gamma, w\delta)$, where $\alpha, \beta, \gamma, \delta \in F_0$ and $w \notin F_0$. Each point $(w\gamma, w\delta)$ can be expressed in terms of z (used above to express the action of ψ) as $((z+\xi)\gamma', (z+\xi)\delta')$, where

$\xi, \gamma', \delta' \in F_0$. Furthermore, if an element of G fixes a point $\big((z+\xi)\gamma, (z+\xi)\delta\big) \neq (0, 0)$, then it fixes all points $\big((z+\xi)\gamma', (z+\xi)\delta'\big)$, where γ', δ' are arbitrary in F_0 and ξ is fixed. Clearly, the point sets

$$\pi(\infty) = \{(\alpha, \beta) \mid \alpha, \beta \in F_0\} \cup \{Q_i \mid i \neq 0\} ,$$

$$\pi(\xi) = \{\big((z+\xi)\gamma, (z+\xi)\delta\big) \mid \gamma, \delta \in F_0\} \cup \{Q_i \mid i \neq 0\} ,$$

constitute a collection of $p^t + 1$ Baer subplanes of π which form the lines in $\overline{\pi}_0$ through $(0, 0)$ (in fact, $\pi(\infty)$ is the line $x = 0$ in $\overline{\pi}$ and $\pi(\xi)$ the line $y = x\xi$ in $\overline{\pi}$). The following table shows the composition of various subgroups of G .

Group	Coefficients
R_1	$d_2 = 0$, $e_2 = 1$
R_2	$d_1 = 1$, $e_1 = 0$
G_1	$d_2 = 0$
G_2	$e_1 = 0$
$G_{12} = G_1 \cap G_2$	$d_2 = e_1 = 0$
$R_1 \cap G_2$	$d_2 = e_1 = 0$, $e_2 = 1$
K	$d_2 = e_1 = 0$, $d_1 = e_2$.

In the table R_1, R_2 are the pointwise stabilizers of $\pi(\infty)$ and $\pi(0)$ respectively, G_1 and G_2 are the stabilizers of $\pi(\infty)$ and $\pi(0)$ respectively, and K is the kernel of the natural homomorphism of G onto G^Ω , where $\Omega = \{\pi(\eta) \mid \eta \in F_0 \text{ or } \eta = \infty\}$.

Notice that K is the group of $(0, l_\infty)$-homologies of the Hall plane and so is cyclic of order $p^t - 1$. Also, $K \cap R_i = 1$ and so $R_i^\Omega \cong R_i$. It is easy to verify that $\left|R_1^\Omega \cap R_2^\Omega\right| = p^t - 1$ and that $G^\Omega \cong G/K \cong \mathrm{PGL}(2, p^t)$ in its natural representation. This leads to a question.

QUESTION 1. *Let* G *be a partially transitive group of type* $(4, m)$ *, and let* Ω *be the set of* m *Baer subplanes whose points constitute the lines* K_i *,* $i = 1, \ldots, m$ *. Does it then follow that* $m = p^t + 1$ *for some prime* p *, and* $G^\Omega \cong \mathrm{PGL}(2, p^t)$ *in its natural representation?*

It is interesting to note that the Hughes plane π^* of order p^{2t} has a desarguesian subplane π_0^* of order p^t, and a group G^* acting analogously on π^* as $\bar{G}$ does on $\bar{\pi}$. The planes derived from the Hughes planes (the Ostrom-Rosati planes) will be of type $(4, p^t+1)$ and Question 1 can be answered affirmatively for these planes also.

2. Type $(5, m)$. In this section we shall consider the action of the collineation group $G = T(Y)G(\pi_0)$ of a generalized Hall plane which was introduced in the proof of Theorem 1 (ii). We shall simply state a few results which the reader can easily verify for himself.

The Baer subplanes whose points constitute the lines K_i, $i = 1, \ldots, p^t$ are given by

$$\pi(0) = \{(\alpha, \beta) \mid \alpha, \beta \in F_0\} \cup \{Q_i \mid i = 0, 1, \ldots, p^t\}$$

and

$$\pi(\xi) = \{(\alpha, z\xi+\beta) \mid \alpha, \beta \in F_0\} \cup \{Q_i \mid i = 0, 1, \ldots, p^t\}$$

where $\xi \neq 0$, $z \notin F_0$. The stabilizer G_1 of $\pi(0)$ is the semi-direct product of the group of translations T' of the form $(x, y) \to (x, y+\beta)$, $\beta \in F_0$, and the pointwise stabilizer R_1 of $\pi(0)$ (which is $G(\pi_0)$). Hence $|G_1| = p^t(p^{2t}-p^t) = m^2(m-1)$. Denoting the stabilizer of $\pi(1)$ by G_2 we have that $G_{12} = G_1 \cap G_2$ is the semi-direct product of the translation group T' and the subgroup Φ of $G(\pi_0)$ consisting of collineations $(x, y) \to (\phi(x), \phi(y))$, where $\phi(z\alpha+\beta) = z\alpha + s\alpha + \beta$, $s \in F_0$. Hence $|G_{12}| = p^{2t} = m^2$. Also the group $R_1 \cap G_2 = \Phi$ and so $|R_1 \cap G_2| = p^t = m$.

Finally, the kernel K of the natural homomorphism of G onto G^Ω is, in fact, G_{12} and so $|K| = p^{2t} = m^2$. It follows that $|G^\Omega| = p^t(p^t-1)$ and so G^Ω is a sharply doubly transitive group on Ω. It is easily seen that G^Ω is isomorphic to the group of linear substitutions on $F_0 = GF(p^t)$, and this brings us to a second question.

QUESTION 2. *Let G be a partially transitive group of type $(5, m)$ and let Ω be the set of m Baer subplanes whose points constitute the lines K_i, $i = 1, \ldots, m$. Does it then follow that $m = p^t$ for some prime p and G^Ω is* *isomorphic to the group of linear substitutions on* $GF(p^t)$?

3. Type $(6, m)$. A discussion of partially transitive planes of type $(6, m)$ appears in [19], where it is proved that, if π is a projective plane of Hughes type $(6, p)$, where p is a prime, then $p \leq 3$. At this point we simply take the opportunity of asking another question.

QUESTION 3. *If* π *is a projective plane of Hughes type* $(6, m)$ *, then is* $m \leq 3$?

The results of the following sections arose out of an attempt to answer the three questions we have posed in this section.

6. Investigation of the collineation group G

When trying to determine the collineation group G acting doubly transitively on the set Ω of Baer subplanes, the main problem is that we have no obvious control of the kernel K of the action.

Although the groups R_i and R_j intersect trivially for $i \neq j$, the corresponding subgroups $R_i^\Omega = R_iK/K$ and $R_j^\Omega = R_jK/K$ of $G^\Omega = G/K$ need not intersect in a group acting trivially on Ω . Lemmas 6.1 and 6.3 describe the situation. Now whether or not the groups $K \cap R_i$ are trivial is important and Theorem 6.5 gives the values of m and describes the group G for each type of plane for which $K \cap R_i$ is nontrivial. We use this result to derive Theorems 6.6 and 6.7 which are our main characterisation results. The results for planes of types $(5, m)$ and $(6, m)$ offer some evidence for an affirmative answer to Questions 2 and 3 of §5. However for planes of type $(4, m)$, as well as the case $G/K \geq PSL(2, m-1)$, we also have the possibility that m is a power of a prime p and G/K has a regular normal p-subgroup. The only cases where this is known to occur are $m = 2, 3, 4$; so we ask

QUESTION 4. *Do planes of type* $(4, p^s)$ *exist where* p *is a prime,* $s \geq 1$ *,* $p^s > 4$ *, for which the group* $G^\Omega = G/K$ *has a normal elementary abelian* p*-subgroup of order* p^s *acting regularly on* Ω ?

The results of the final §7 give some information about the action on Ω of Sylow p-subgroups Q of G_{ij} for primes p dividing $|G_{ij}|$. In particular we obtain inequalities involving $|Q|$, $|Q \cap K|$, and $|\mathrm{fix}_\Omega Q|$ which give exact values for these quantities when $|Q|$ is as small as possible; that is, p^2 . We obtain the result of Rahilly and Searby [19] for planes of type $(6, m)$, m a prime, as Corollary 7.2 *(c)*.

Before beginning we remark that

(i) the groups $\{R_i \cap G_j \mid i \neq j,\ i,\ j = 1, \ldots, |\Omega|\}$ are all conjugate in G ; and

(ii) the groups $\{K \cap R_i \mid i = 1, \ldots, |\Omega|\}$ are all conjugate in G .

LEMMA 6.1. $G_{ij} = (R_i \cap G_j) \times (R_j \cap G_i)$.

Proof. Since $R_i \cap G_j$, and $R_j \cap G_i$ are normal subgroups of G_{ij} which intersect trivially, it follows that the group they generate is the direct product $X = (R_i \cap G_j) \times (R_j \cap G_i) \leq G_{ij}$. Then, as in all cases $|G_{ij}| = |R_i \cap G_j|^2$, it follows that $X = G_{ij}$ and the lemma is proved.

COROLLARY 6.2. *The group* G_i *is the semi-direct product of* R_i *and* $R_j \cap G_i$ *for any* $j \neq i$ *(that is* $R_i \trianglelefteq G_i$, $R_j \cap G_i \leq G_i$, $R_i \cap (R_j \cap G_i) = 1$, *and* $G_i = R_i(R_j \cap G_i)$ *).*

The proof is obvious.

Now let π_{ij}, π_{ji} be the projections of G_{ij} onto $R_i \cap G_j, R_j \cap G_i$ respectively. Let $K_{ij} = \pi_{ij}(K)$, $K_{ji} = \pi_{ji}(K)$, and since K is normal in G_{ij} then K_{ij}, K_{ji} are normal subgroups of $R_i \cap G_j$, $R_j \cap G_i$ respectively.

LEMMA 6.3. *If* $i \neq j$ *, then with the above notation,* $KR_i \cap KR_j = K_{ij} \times K_{ji}$ *, so that* $|KR_i \cap KR_j| = |K|^2/|K \cap R_i|^2$ *, and* $|K| \geq |K \cap R_i|^2$ *. Moreover* $KR_i \cap KR_j = K$ *if and only if* $|K| = |K \cap R_i|^2$ *, and this happens if and only if either* $K = 1$ *, or* $K \cap R_i \neq 1$ *and* $K = (K \cap R_i) \times (K \cap R_j)$ *.*

Proof. Clearly $\langle K_{ij}, K_{ji} \rangle = K_{ij} \times K_{ji}$, and $K_{ij} \leq R_i \cap G_j < KR_i$. If $k \in K_{ij}$ then there is an $h \in K_{ji}$ such that $kh \in K$, and so
$k = khh^{-1} \in KK_{ji} < KR_j$. Hence $K_{ij} < KR_i \cap KR_j$, and similarly K_{ji} and hence $K_{ij} \times K_{ji}$ are subgroups of $KR_i \cap KR_j$.

Now we shall show that $KR_i \cap G_j \leq (R_i \cap G_j) \times K_{ji}$, and $G_i \cap KR_j \leq K_{ij} \times (R_j \cap G_i)$ and therefore

$$KR_i \cap KR_j \leq \left((R_i \cap G_j) \times K_{ji}\right) \cap \left(K_{ij} \times (R_j \cap G_i)\right) = K_{ij} \times K_{ji} ,$$

and equality follows. It is sufficient to show that $KR_i \cap G_j \leq (R_i \cap G_j) \times K_{ji}$.

Let $g \in KR_i \cap G_j$. Then $g = hk = hk_1k_2$, where $h \in R_i$, $k_1 \in K_{ij} \le R_i \cap G_j$, $k_2 \in K_{ji}$, and $k = k_1k_2 \in K$. Further since $h = gk^{-1} \in G_j$, then $hk_1 \in R_i \cap G_j$. Thus $g \in (R_i \cap G_j) \times K_{ji}$ and the proof that $KR_i \cap KR_j = K_{ij} \times K_{ji}$ is complete.

The kernels of π_{ij}, π_{ji} are $R_j \cap G_i$ and $R_i \cap G_j$ respectively, and hence $|K_{ij}| = |K|/|K \cap (R_j \cap G_i)| = |K|/|K \cap R_j|$, and $|K_{ji}| = |K|/|K \cap R_i|$. Since $K \cap R_i$ and $K \cap R_j$ are conjugate it follows that

$$|KR_i \cap KR_j| = |K_{ij}||K_{ji}| = |K|^2/|K \cap R_i|^2 .$$

Further since $K \le KR_i \cap KR_j$ it follows that $|K| \ge |K \cap R_i|^2$, and that $K = KR_i \cap KR_j$ if and only if $|K| = |K \cap R_i|^2$. Under this condition either $K \cap R_i = 1$ so that $K = 1$, or $K \cap R_i \ne 1$. In the latter case $K \cap R_i$, $K \cap R_j$ are nontrivial normal subgroups of K with trivial intersection. Thus clearly $K = (K \cap R_i) \times (K \cap R_j)$. This completes the proof.

We shall use the following result in the proofs of later theorems.

LEMMA 6.4. *If the group G^Ω satisfies $\mathrm{PSL}(t, q) \le G^\Omega \le \mathrm{P\Gamma L}(t, q)$ for some $t \ge 3$ and prime power q, in its natural representation on $|\Omega| = (q^t-1)/(q-1)$ points or hyperplanes, then the plane is of type $(6, 2)$ or $(6, 3)$.*

Proof. Suppose that $\mathrm{PSL}(t, q) \le G/K \le \mathrm{P\Gamma L}(t, q)$ and that $|\Omega| = (q^t-1)/(q-1)$ for some $t \ge 3$. Let

$$x = |\mathrm{PSL}(t, q)|/\big(|\Omega|(|\Omega|-1)\big) = \begin{cases} q^{\binom{t}{2}-1}(q^{t-2}-1) \ldots (q^2-1)(q-1)^2/(t, q-1), & \text{if } t \ge 4, \\ q^2(q-1)^2/(3, q-1), & \text{if } t = 3. \end{cases}$$

Then certainly x divides $|G|/\big(|\Omega|(|\Omega|-1)\big)$.

In type $(4, m)$, we have $|\Omega| = m$ and $|G|/m(m-1) = (m-2)^2$, and since $m = |\Omega| \equiv 1 \pmod q$ it follows that $(m-2)^2 \equiv 1 \pmod q$, and so is not divisible by x for any t, q. Hence planes of type $(4, m)$ do not arise.

In type $(5, m)$, we have $|\Omega| = m$ and $|G|/m(m-1) = m^2$. As above

$m^2 \equiv 1 \pmod q$ and so is not divisible by x for any t, q. Hence planes of type $(5, m)$ do not arise.

Thus the plane is of type $(6, m)$, $|\Omega| = 1 + m + m^2$, and $|G|/(1+m+m^2)(m+m^2) = m^2(m-1)^2$. Now $m^2 < m^2 + m = q(q^{t-1}-1)/(q-1) < 2q^{t-1}$, and so $m^2(m-1)^2 < 4q^{2(t-1)}$. On the other hand if $t \geq 4$ it is easy to check that $x > 4q^{2(t-1)}$ unless $t = 4$ and $q = 2$, but in this case $|\Omega| = 15 = 1 + m + m^2$ which is impossible. Thus $t = 3$ and $m = q$. Moreover since $|G|/|K|(1+m+m^2)(m+m^2) = q^2(q-1)^2/|K|$ is divisible by $x = q^2(q-1)^2/(3, q-1)$ it follows that $|K|$ is 1 or $(3, q-1)$, and that

$$|(G/K) : \mathrm{PSL}(3, q)| = (3, q-1)/|K| = \delta$$

is 1 unless $q - 1$ is divisible by 3 and K is trivial, and then δ is 3. Set $\overline{G} = \mathrm{PSL}(3, q)$, $\overline{R}_i = (R_iK/K) \cap \overline{G}$, $\overline{G}_i = (G_i/K) \cap \overline{G}$ and so on. Then as $\overline{G}$ is 2-transitive on Ω, $|(G_{ij}/K) : \overline{G}_{ij}| = \delta$ and it is easy to see from Lemma 6.1 that $|(R_iK/K) : \overline{R}_i| = \delta$.

The subgroup T of $\overline{G}_i$ generated by the transvections in $\overline{G}_i$ is a normal elementary abelian subgroup of $\overline{G}_i$ of order q^2 and all its orbits in $\Omega - \{i\}$ have length q. For $j \neq i$, the group $T \cap \overline{G}_j$ has order q and fixes precisely the points on the line through i and j. Moreover each nontrivial element of T fixes pointwise precisely one of the $q + 1$ lines containing i.

Suppose that $T_i = T \cap \overline{R}_i$ is nontrivial. Then since T_i is normal in $\overline{G}_i$, all T_i-orbits in $\Omega - \{i\}$ have the same length, and since T_i is normal in T, all have length $p^t \leq q$, where q is a power of the prime p. Since any nontrivial element of T fixes points of $\Omega - \{i\}$, then T_i does not act semiregularly on $\Omega - \{i\}$. It follows from O'Nan [13, Lemma 4.8] that Ω is a projective space of a field of p^t elements of dimension $r - 1$ for some $r \geq 3$. Hence

$$|\Omega| = 1 + q + q^2 = (p^{tr}-1)/(p^t-1) \equiv p^t + 1 \pmod{p^{t+1}},$$

and as q is a power of p, $q \geq p^t$, it follows that $q = p^t$. Now by O'Nan [13, Lemma 4.6] T_i has order at least $p^{2t} = q^2$ and hence $T_i = T \leq \overline{R}_i$. Thus we have shown that either $T \cap \overline{R}_i = 1$ or $T \leq \overline{R}_i$.

Now consider Ω homogeneously coordinatized by $\mathrm{GF}(q)$ in such a way that i

has coordinates $\begin{pmatrix}0\\0\\1\end{pmatrix}$. The elements of the group $\overline{G}_i$ can be represented (uniquely) by nonsingular 3×3 matrices of the form

$$M = \begin{pmatrix} a & b & 0 \\ c & d & 0 \\ e & f & 1 \end{pmatrix}$$

over $GF(q)$. Then the homomorphism $\chi : \overline{G}_i \to GL(2, q)$ defined by

$$\chi(M) = \begin{pmatrix} a & b \\ c & d \end{pmatrix}$$

has kernel consisting of matrices of the form

$$\begin{pmatrix} 1 & 0 & 0 \\ 0 & 1 & 0 \\ e & f & 1 \end{pmatrix} .$$

This is precisely the group T . Now $\chi(\overline{G}_i)$ is a subgroup of $GL(2, q)$ containing $SL(2, q)$ since $\overline{G} = PSL(3, q)$, and containing a normal subgroup $\chi(\overline{R}_i) \simeq \overline{R}_i/(\overline{R}_i \cap T)$ of order $q^2(q^2-1)/\delta$, $(q^2-1)/\delta$ as $T \cap \overline{R}_i = 1$, $T \leq \overline{R}_i$ respectively. Now for $q \geq 4$, any normal subgroup of $\chi(\overline{G}_i)$ either is central or has an insoluble composition factor isomorphic to $PSL(2, q)$. Since the centre of $\chi(\overline{G}_i)$ has order dividing $q - 1$ and since the order of $\chi(\overline{R}_i)$ is divisible by $q + 1$, it follows that $\chi(\overline{R}_i)$ has $PSL(2, q)$ as a composition factor and hence has order divisible by $q(q^2-1)/(2, q-1)$. Therefore $T \cap \overline{R}_i = 1$ and $|\chi(\overline{R}_i)| = q^2(q^2-1)/\delta$ is divisible by q^2 , a contradiction since $|GL(2, q)|$ is not divisible by q^2 . Thus we conclude that $q \leq 3$, and the proof of Lemma 6.4 is complete.

We can now find restrictions on m , for each type of plane, implied by the condition that $K \cap R_i$ is nontrivial.

THEOREM 6.5. *Suppose that* $|K \cap R_i| = y > 1$. *Then*

(a) planes of type $(6, m)$ *do not occur for any* m *, and*

(b) for planes of types $(4, m)$ *and* $(5, m)$ *,* m *takes the values* $p^s + 2 = q^t$ *, and* p^s *respectively, where* p*,* q *are primes and* s *and* t *are positive integers.*

Further K *is elementary abelian of order* $p^s y$ *, where* $y = |K \cap R_i|$ *divides* p^s *, and* G/K *has an elementary abelian normal subgroup of order* m *acting regularly on* Ω . *If* m *is even then* G *is soluble.* (*Also in type* $(4, m)$ *the integer* $s \geq 2$ *,* *and in type* $(5, m)$, G_{ij} *fixes* p^a *points of* Ω *for some* a *dividing* s .)

Proof. Suppose that $|R_i \cap K| = y > 1$ and recall that all the $R_i \cap K$ are conjugate in G . Then as K is normal in G and as G is transitive on $l - (l \cap \pi')$ for l a line of π' , it follows that K has all orbits in $l - \pi'$ of the same length z , and as K fixes $(l-\pi') \cap \pi_i$ setwise for all $\pi_i \in \Omega$, then

(1) $$z \text{ divides } \xi = |(l-\pi') \cap \pi_i|$$

where $\xi = m-2, m, m(m-1)$ in types $(4, m)$, $(5, m)$, $(6, m)$ respectively. Now $R_i \cap K$ is the stabiliser in K of a point of $(l-\pi') \cap \pi_i$, and so $|K : R_i \cap K| = z$. Thus

(2) $$|K| = zy .$$

Let p be a prime dividing y , and let C_i be a subgroup of $R_i \cap K$ of order p , for each i . Now if $i \neq j$ then $R_i \cap K$, $R_j \cap K$ are normal subgroups of K which intersect trivially and hence centralise each other. It follows that the group $C = \langle C_i \mid \pi_i \in \Omega\rangle$ is an elementary abelian p-subgroup of K . Then $|C| = p^r$ for some $r \geq 2$, and C has $(p^r-1)/(p-1)$ distinct subgroups of order p . Then since $\{C_i \mid \pi_i \in \Omega\}$ is a set of $|\Omega|$ distinct subgroups of C of order p we conclude that $|\Omega| \leq (p^r-1)/(p-1)$. Let s be the integer satisfying

(3) $$1 \leq s < r , \quad |\Omega| \leq (p^{s+1}-1)/(p-1) , \quad |\Omega| > (p^s-1)/(p-1) .$$

Let C' be a subgroup of C of order p^s . Then C' has $(p^s-1)/(p-1)$, that is, less than $|\Omega|$, distinct subgroups of order p . We note that any nontrivial element of G fixes a point of a set $(l-\pi') \cap \pi_i$ for at most one value of i , for if the element g of G fixes points of both $(l-\pi') \cap \pi_i$ and $(l-\pi') \cap \pi_j$ for some $i \neq j$, then $g \in R_i \cap R_j = 1$, that is $g = 1$. It follows, since C' has less than $|\Omega|$ subgroups of order p that there is some k such that no nontrivial element of C' fixes a point of $(l-\pi') \cap \pi_k$. Since $C' \leq K$, then C' fixes $(l-\pi') \cap \pi_k$ setwise, and so each orbit of C' in $(l-\pi') \cap \pi_k$ has length $|C'| = p^s$. Thus p^s divides ξ where ξ is as in (1). Suppose that $p^s \leq \xi/2$. Then by (3) we have $\xi \geq 2p^s > (p^{s+1}-1)/(p-1) \geq |\Omega|$. This is a contradiction as in all cases $\xi \leq |\Omega|$. Hence $\xi = p^s$.

For planes of type $(6, m)$ we have $p^s = m(m-1)$ which is impossible unless

$m = 2 = p^s$. However in this case $|\Omega| = 7$ whereas by (3), $|\Omega| \le 3$. Thus planes of type $(6, m)$ do not arise.

In type $(4, m)$ and $(5, m)$ we have $p^s = \xi = m-2, m$ respectively. Now $|\bigcup_{1 \le i \le m} (R_i \cap K)| = 1 + m(y-1) \le yz = |K|$, that is, $m - 1 \ge y(m-z)$. If $z \le \frac{1}{2}\xi$ then we have $m - 1 \ge y(m-\xi/2) \ge ym/2$, that is, $y < 2$, contradiction. Thus as z divides ξ by (1), we have $z = \xi = p^s$. Hence C' acts transitively and regularly on $(l-\pi') \cap \pi_k$. Clearly $K = (K \cap R_k)C'$, and $(K \cap R_k) \cap C' = 1$, since the kernel of the action of K on $(l-\pi') \cap \pi_k$ is $K \cap R_k$. It follows that the constituent of K on this set, namely $K/(K \cap R_k) \simeq C'$, is an elementary abelian p-group, and (since K is isomorphic to a subgroup of $\prod_{1 \le i \le m} K/(K \cap R_i)$), then K is an elementary abelian p-group. Further, since

$$|KR_i| = |K||R_i|/|K \cap R_i| = p^{2s}(m-1) = |G_i| ,$$

it follows that $G_i = KR_i$, and so by Lemma 6.3, $G_{ij} = K_{ij} \times K_{ji}$. Since K_{ij} and K_{ji} are homomorphic images of K , both are elementary abelian p-groups and hence G_{ij} is an elementary abelian p-group. Moreover $|G_{ij}/K| = p^s/y$ so that y divides p^s . It follows from Aschbacher [1] that one of the following holds:

(a) G/K has a regular normal subgroup;

(b) $G/K = \mathrm{PSL}(3, 2)$, $m = 7$;

(c) $G/K = R(3)$, the smallest Ree group, and $m = 28$;

(d) G/K contains a unique minimal normal subgroup such that $G/K \le \mathrm{Aut}\ N$, and N is $\mathrm{PSL}(2, q)$, $\mathrm{PSU}(3, q)$, $Sz(q)$, or a group of Ree type $R(q)$, where $m = q^\delta + 1$ and δ is 1, 3, 2, 3 respectively.

(Clearly $q \ge 3$ since for $q = 2$, G/K has a regular normal subgroup.)

Clearly (b) does not arise by Lemma 6.4. Next we consider cases (c) and (d) with $N = R(3)$, $m = 3^3 + 1$, in case (c). In all cases the two point stabiliser N_{ij} has order divisible by $q - 1$, and so $q - 1$ divides $|G_{ij}/K| = p^s/y$. Thus $q - 1 = p^t$ for some integer t , $1 \le t < s$. Hence $m = q^\delta + 1 = (q^\delta - 1) + 2 \equiv 2 \pmod{p^t}$ and so the plane is of type $(4, m)$ with $m = p^s + 2$, $q^\delta - 1 = p^s$, unless $p = p^t = q - 1 = 2$. In the latter case $m = 4, 10, 28$ as $\delta = 1, 2, 3$ respectively so the plane is of type $(4, m)$ unless $m = 4$ and in that case, (a) holds. Thus suppose the plane is of type $(4, m)$ with $q^{\delta} - 1 = p^{s}$. It is easy to show for $\delta = 2$ or 3 and $q > 2$ that $q^{\delta} - 1$ is a prime power if and only if $\delta = 2$ and $q = 3$. However if $\delta = 2$, then $N = Sz(q)$ and q is a power of 2 . Thus $\delta = 1$, and $p^{s} = q - 1 = p^{t}$, a contradiction since $t < s$.

Thus G/K has a regular elementary abelian normal subgroup N/K of order $m = q^{t}$ for some prime q , integer t . In type $(4, m)$, $q^{t} = p^{s} + 2$, and in type $(5, m)$, $q^{t} = p^{s}$. If m is even then either $m = 4$ and G is soluble, or the plane is of type $(5, m)$ and $m = 2^{s}$. In Ashbacher [1] it is shown that for m even, either G is soluble or $s = 2s'$ is even and G/K is the group $\mathrm{PSL}(2, 2^{s'})$ acting on a group of order 2^{s} . However $2^{s}|\mathrm{PSL}(2, 2^{s'})| = 2^{3s'}(2^{s}-1)$, while $|G/K| = 2^{3s'}(2^{s}-1)/y$, and since $y > 1$ the latter case is impossible. Hence G is soluble if m is even.

Finally:

(i) If $s = 1$ then $|K| = py = p^{2}$ and K has $p + 1$ distinct subgroups of order p . Since $\{K \cap R_{i}\}$ is a set of m distinct subgroups of order p , then $m \le p + 1$ and hence the plane is of type $(5, m)$.

(ii) In type $(5, m)$, if $k = |\mathrm{fix}_{\Omega}\, G_{ij}|$ then since $|G_{ij}|$ and $|\Omega|$ are powers of p , it follows that p divides k . Suppose $2 < k < p^{s}$, that is, $G_{ij} > K$. Then by Witt's Lemma (Witt [23]), $N_{G}(G_{ij})$ is sharply 2-transitive on $\mathrm{fix}_{\Omega}\, G_{ij}$. It follows that k is a prime power. Hence $k = p^{a}$. The sets $\{\mathrm{fix}\, G_{ij} \mid i \ne j\}$ form the blocks of a block design with $\lambda = 1$, with $m = p^{s}$ points, and say b blocks. From the equation (obtained by counting incident block-point pairs), $p^{s}(p^{s}-1)\lambda = bp^{a}(p^{a}-1)$, it follows that $p^{a} - 1$ divides $p^{s} - 1$ and hence that a divides s . (For more details see Kantor [10], §6A.) This completes the proof of Theorem 6.5.

We can now state and prove the main results characterising G , Theorems 6.6 and 6.7.

THEOREM 6.6. *Suppose the plane is of type* $(4, m)$ *with* m *odd or* $m \equiv 0, 4$, *or* $6 \pmod 8$, *or of type* $(5, m)$ *with* m *odd or* $m \equiv 2 \pmod 4$. *Then either (a) or (b) holds.*

(a) $G^{\Omega} = G/K$ *has a regular elementary abelian normal subgroup of order* $m = p^{s}$ *for some prime* p *and positive integer* s . *If* p *is odd, the Sylow* 2*-subgroups*

of G *are cyclic or generalised quaternion. If* p *is* 2 *then for type* $(5, m)$ *,* m *is* 2 .

(b) The plane is of type $(4, m)$ *, where* $m = 1 + p^s$ *for some prime* p *and positive integer* s *, and* $\mathrm{PSL}(2, p^s) \le G/K \le \mathrm{P\Gamma L}(2, p^s)$ *. In particular if* $G/K = \mathrm{PGL}(2, p^s)$ *then* K *is cyclic of order* $p^s - 1$ *, but in any case* $|K|$ *is a multiple of* $(p^s-1)/(s, p^s-1)$.

Proof. (i) We consider the case m odd first, and apply a result of Bender. Since $|G_{ij}|$ is odd it follows from Bender [3, Theorem 3] that either

(a) G/K has a regular normal subgroup, or

(b) G/K has a normal subgroup N such that $G \le \mathrm{Aut}\, N$ and N is $\mathrm{PSL}(2, q)$, $\mathrm{PSU}(3, q)$ or $Sz(q)$ where q is a power of 2 , $q \ge 4$.

In case (a) it follows from Bender [3, Theorem 1] that Sylow 2-subgroups of G/K are cyclic or generalised quaternion. In case (b) it can be shown as in the proof of Theorem 6.5 that $N = \mathrm{PSL}(2, q)$ with $m = q + 1 = 2^s + 1$ for some $s \ge 2$. Now $|\mathrm{P\Gamma L}(2, 2^s)| = m(m-1)(m-2)s$, and it follows that $|G_{ij}/K| = (m-2)t$ for some t dividing s , and $|K| = (m-2)/t$. Hence t divides $(s, 2^s-1)$ and $|K|$ is a multiple of $(2^s-1)/(s, 2^s-1)$. In particular if $G = \mathrm{PSL}(2, q) = \mathrm{PGL}(2, q)$ then $t = 1$ and by Lemma 6.3 and Theorem 6.5, $G_{ij} = K_{ij} \times K_{ji}$, where K_{ij} and K_{ji} are isomorphic to K . Since G_{ij}/K is cyclic of order $m - 2$, and since $G_{ij}/K = K_{ij}K/K \simeq K_{ij} \simeq K$ it follows that K is cyclic of order $m - 2$.

(ii) Now suppose that the plane is of type $(4, m)$ with $m \equiv 0 \pmod 4$ or of type $(5, m)$ with $m \equiv 2 \pmod 4$. Then $|R_j \cap G_i|$ and $|R_i|$ are both twice an odd number and hence $R_j \cap G_i$ and R_i are soluble (for by Burnside's Transfer Theorem $R_j \cap G_i$ and R_i have normal odd-order subgroups of index 2 which by the "Odd-order" theorem of Feit and Thompson are soluble). Thus since R_i is normal in G_i and $G_i/R_i \simeq R_j \cap G_i$ by Lemma 6.2, it follows that G_i is soluble. Hence by a theorem of Holt [4], G/K has a normal subgroup N , where either

(a) N is an elementary abelian 2-group of order $m = 2^s$, or

(b) N is isomorphic to $\mathrm{PSL}(2, q)$, $\mathrm{PSU}(3, q)$ or a group $R(q)$ of Ree type, and $m = q + 1$ where δ is 1, 3, 3 respectively.

In case (a), for planes of type $(5, m)$, it follows since $m \equiv 2 \pmod 4$ that $m = 2$. Moreover if $K \cap R_i$ is nontrivial then from Theorem 6.5 it follows that case (a) occurs. So suppose that case (b) occurs. Then $K \cap R_i$ is trivial and by Lemma 6.3, $|G_{ij}/K|$ is divisible by $|K|$. Hence $|G_{ij}/K|$ is divisible by $m-2$, m for planes of type $(4, m)$, $(5, m)$ respectively.

If $N = \mathrm{PSU}(3, q)$ then $(q^2-1)/(3, q-1)$ divides $|G_{ij}/K|$, and hence divides $|G_{ij}| = (q^3-1)^2$, $(q^3+1)^2$ for types $(4, m)$, $(5, m)$ respectively. This is impossible for types $(4, m)$ for any q, and for type $(5, m)$, q must be 2 or 3, but since $m \equiv 2 \pmod 4$ neither is permissible.

Supoose that N is $R(q)$, a group of Ree type, where $q = 3^s$ for some $s \geq 1$. If $s > 1$ then by Ree [20, Theorem 9.1], the outer automorphism group of N is isomorphic to the group of field automorphisms and so $|G/K|$ divides $(q^3+1)q^3(q-1)s$, that is, $|G_{ij}/K|$ divides $(q-1)s$. However we remarked above that $|G_{ij}/K|$ was divisible by $m-2$ or m, and $m-2 = p^{3s}-1 > (p-1)s$, so this is impossible. Hence $m = 28$, but in this case $R(3)$ has a normal subgroup $\mathrm{PSL}(2, 8)$ of index 3, and $\mathrm{Aut}\, R(3) = R(3)$. Since $m-2 = 26$ or $m = 28$ divides $|G_{ij}/K|$, this case is clearly impossible.

Hence $N = \mathrm{PSL}(2, q)$ and $m = q+1$. For type $(5, m)$ then $(q-1)/2$ divides $|G_{ij}| = (q+1)^2$ and hence q is 3, 5, or 9. Since $m \equiv 2\ (4)$ then q is 5 or 9. However as 3 or 5 respectively divides $|G_{ij}/K|$ it follows that G_{ij}/K contains an element of order 3, 5, and degree 3, 5, and hence by [22, 13.10], $G/K \geq A_m$, a contradiction. Thus the plane is of type $(4, m)$; the rest of (b) follows as in the proof of the case m odd.

(iii) Finally suppose that the plane is of type $(4, m)$ with $m \equiv 6 \pmod 8$. Then a Sylow 2-subgroup of $R_i \cap G_j$ has order 4 and hence is abelian. From Lemma 6.1 then, a Sylow 2-subgroup of G_{ij} is abelian. Thus since $m \equiv 2 \pmod 4$ it follows from Aschbacher [2, Theorem 2] that G/K has a normal subgroup N such that $G \leq \mathrm{Aut}\, N$ and N is isomorphic to A_6, $\mathrm{PSL}(2, q)$ or $\mathrm{PSU}(3, q)$ with $m = 6$, $q+1$, or q^3+1 respectively. As above, the groups $\mathrm{PSU}(3, q)$ do not arise, and clearly neither does A_6. Thus N is $\mathrm{PSL}(2, q)$ and part (b) holds as above.

REMARK. Clearly the conclusions of Theorem 6.6 hold if the plane is of type $(4, m)$, $m \equiv 2 \pmod 8$, and if the Sylow 2-subgroups of $R_i \cap G_j$ are abelian.

THEOREM 6.7. *Suppose that* $|K| \le |K \cap R_i|^2$. *Then either*

(a) the plane is of type $(6, m)$ *where* m *is* 2 *or* 3 , K *is trivial, and* $\mathrm{PSL}(3, m) \le G/K \le \mathrm{P\Gamma L}(3, m)$; *or*

(b) the plane is of type $(4, m)$ *or* $(5, m)$ *and* $m = p^s + 2 = q^t$, $m = p^s$ *respectively, where* p, q *are primes and* s, t *are positive integers. Further* G/K *is sharply* 2-*transitive on* Ω , *and* K *is an elementary abelian* p-*group of order* p^{2s} .

Note that the condition $|K| \le |K \cap R_i|^2$ is equivalent by Lemma 6.3 to $KR_i \cap KR_j = K$, for $i \ne j$; that is $R_i^{\Omega} \cap R_j^{\Omega} = 1$.

Proof. As we noted above the condition $|K| \le |K \cap R_i|^2$ is equivalent to $R_i^{\Omega} \cap R_j^{\Omega} = 1$ for $i \ne j$; that is R_i^{Ω} is a T.I. set. If R_i is not semiregular on $\Omega - \{i\}$, then by O'Nan [14, Theorem A], $\mathrm{PSL}(t, q) \le G/K \le \mathrm{P\Gamma L}(t, q)$ for some $t \ge 3$ and prime power q . By Lemma 6.4 the plane is of type $(6, 2)$ or $(6, 3)$, and $t = 3$, $q = m = 2$ or 3 . Also by Theorem 6.5, K is trivial.

On the other hand, if R_i acts semiregularly on $\Omega - \{i\}$, then as $|R_i| > |\Omega|$ in all cases, it follows that $K \cap R_i$ is nontrivial. It follows from Theorem 6.5 that the plane is of type $(4, m)$ or $(5, m)$ with $m = p^s + 2 = q^t$ or $m = p^s$ respectively, where p and q are primes, s and t are positive integers; also that K is an elementary abelian p-group of order $p^s|K \cap R_i|$. Since by Lemma 6.3, $|K| = |K \cap R_i|^2$ it follows that $|K \cap R_i| = p^s$, $|K| = p^{2s}$, and hence that G^{Ω} is sharply 2-transitive.

7.

In this final section we assume that $R_i \cap K$ is trivial. We begin an examination of the Sylow subgroups of G_{ij} and produce some results based on numerical properties of m .

LEMMA 7.1. *Assume that* $R_i \cap K$ *is trivial. Let* p *be a prime dividing* $m-2$, m, $m(m-1)$ *for types* $(4, m)$, $(5, m)$, $(6, m)$ *respectively. Let* P_{ij}, P_{ji} *be Sylow* p-*subgroups of* $R_i \cap G_j$, $R_j \cap G_i$ *respectively, where* $i \ne j$. *Then*

$Q = P_{ij} \times P_{ji}$ *is a Sylow p-subgroup of* G_{ij} . *Moreover*

(1) $1 \le |Q \cap K| \le |Q| - |\text{fix}_\Omega\, Q|\big(|P_{ij}|-1\big)$ *and in particular,*

(2) $2 \le |\text{fix}_\Omega\, Q| \le |P_{ij}| + 1$.

Note that the main value of this result is the bound on $|\text{fix}_\Omega\, Q|$, for by Lemma 6.3 we can deduce, for $K \cap R_i$ trivial, that $|Q \cap K| \le |P_{ij}|$. Using the bound on $|\text{fix}_\Omega Q|$ we deduce the result of Rahilly and Searby [19] as follows.

COROLLARY 7.2. *In type* (6, m) ,

(a) if m *is divisible by a prime* p *to the first power only, then with* Q *as above,* $|\text{fix}_\Omega Q| = p + 1$ *and* $|Q \cap K| = 1$;

(b) if m *is square free then* $|K|$ *is prime to* m ;

(c) (Rahilly and Searby [19]) *if* m *is prime then* $m \le 3$.

Proof of Lemma 7.1. Assume the hypotheses and notation of the statement. By Lemma 6.1, it follows that $Q = P_{ij} \times P_{ji}$ is a Sylow p-subgroup of G_{ij} . Let $\pi_k \in \text{fix}_\Omega\, Q - \{\pi_i\}$, and set $P_{ki} = Q \cap R_k$. Then since Q is a Sylow p-subgroup of G_{ik} , it follows that P_{ki} is a Sylow p-subgroup of $R_k \cap G_i$. Hence $\{P_{ki}, P_{ij} \mid \pi_k \in \text{fix}_\Omega Q - \{\pi_i\}\}$ is a set of $|\text{fix}_\Omega Q|$ normal subgroups of Q , all conjugate in G (since the $R_k \cap G_i$ are conjugate in G for any $k \ne i$) , and intersecting trivially in pairs. Thus any pair centralises each other. Since $R_i \cap K = 1$ for all i it follows that each P_{ki} , and P_{ij} , also intersect K trivially, and so

$$1 \le |Q \cap K| \le \left|Q - \left(\big(\bigcup_k P_{ki}\big) \cup P_{ij}\right)\right| + |\{1\}|$$

$$= |Q| - |\text{fix}_\Omega\, Q|\big(|P_{ij}|-1\big) .$$

Since $|Q| = |P_{ij}|^2$ this inequality yields

$$|\text{fix}_\Omega\, Q| \le \left(|P_{ij}|^2 - 1\right)/\big(|P_{ij}|-1\big) = |P_{ij}| + 1 .$$

Finally since $Q \le G_{ij}$ then Q fixes at least two points of Ω .

Proof of Corollary 7.2 . (a) For planes of type (6, m) suppose that m is divisible by a prime p to the first power only. By Theorem 6.5, $K \cap R_i$ is trivial and the conclusions of Lemma 7.1 hold. Then $|P_{ij}| = |P_{ji}| = p$ and $|Q| = p^2$. Hence $2 \le |\text{fix}_\Omega\, Q| \le p + 1$ and as $|\text{fix}_\Omega\, Q| \equiv |\Omega| \equiv 1 \pmod p$, it follows that $|\mathrm{fix}_\Omega\, Q| = p + 1$. By Lemma 7.1, $|Q \cap K| = 1$ and so $|K|$ is prime to p .

(b) If m is square free then it follows from (a) that $|K|$ is prime to m .

(c) If m is prime, then by (b), m^2 divides $|G_{ij}/K|$ and hence $|G/K|$ is divisible by m^3 exactly. It follows from Tsuzuku [21] that $PSL(3, m) \leq G \leq P\Gamma L(3, m)$ and hence from Lemma 6.5 that $m \leq 3$.

We can obtain a refinement of the bounds for $|\mathrm{fix}_\Omega\, Q|$ of Lemma 7.1 by the technique used in the proof of Theorem 6.5.

LEMMA 7.3. *With the assumptions of Lemma* 7.1, *set* $|P_{ij}| = p^a$, *where* $a \geq 1$. *Then either*

(a) $|\mathrm{fix}_\Omega\, Q| \leq (p^a-1)/(p-1)$, *or*

(b) Q *is elementary abelian.*

Proof. For $k \in \mathrm{fix}_\Omega\, Q$, $Q \cap R_k$ is nontrivial (see Lemma 7.1), and let C_k be a subgroup of $Q \cap R_k$ of order p . Then clearly $C = \langle C_k \mid k \in \mathrm{fix}_\Omega\, Q\rangle$ is an elementary abelian p-group of order say p^r . As C has exactly $(p^r-1)/(p-1)$ subgroups of order p , and as $\{C_k \mid k \in \mathrm{fix}_\Omega\, Q\}$ is a set of $|\mathrm{fix}_\Omega\, Q|$ such subgroups, we have $(p^r-1)/(p-1) \geq |\mathrm{fix}_\Omega\, Q|$. Let s be the integer satisfying $1 \leq s < r$, $(p^s-1)/(p-1) < |\mathrm{fix}_\Omega\, Q| \leq (p^{s+1}-1)/(p-1)$, and let C' be a subgroup of order p^s . Then C' has less than $|\mathrm{fix}_\Omega\, Q|$ subgroups of order p and it follows that there is some $k \in \mathrm{fix}_\Omega\, Q$ such that $C' \cap R_k = 1$. Now $C' \leq G_k$ and so fixes $(l-\pi') \cap \pi_k$ setwise. Hence each orbit of C' in $(l-\pi') \cap \pi_k$ has length p^s . Since the p-part of $|(l-\pi') \cap \pi_k|$ is p^a we conclude that $s \leq a$. Therefore $|\mathrm{fix}_\Omega\, Q| \leq (p^a-1)/(p-1)$ unless $s = a$, so assume that $s = a$. Choose $j \in \mathrm{fix}_\Omega\, Q - \{k\}$. Then $C' \leq G_{jk}$, and $R_j \cap G_k \simeq G_k/R_k$ has a subgroup $C'R_k/R_k \simeq C'$ of order p^a which is elementary abelian. It follows that Sylow p-subgroups of $R_j \cap G_k$, and hence (by Lemma 6.1) of G_{jk} , are elementary abelian. This concludes the proof.

Finally we shall give some miscellaneous results about Sylow subgroups.

LEMMA 7.4. *For planes of type* $(4, m)$, *either the conclusions of* 6.6 *hold, or a Sylow* 2-*subgroup of* G_{ij} *fixes only the two points* i *and* j .

Proof. From 6.6 we may assume that $m \equiv 2 \pmod 8$. Let S be a Sylow 2-subgroup of G_{ij} . Then by Witt [23], $N(S)$ is 2-transitive on fix S and clearly no 2-element in $N(S)$ acting nontrivially on fix S will fix any point of fix S . Hence by Bender [3, Theorem 3], it follows since $|\text{fix } S| \equiv m \equiv 0 \pmod 2$, that $N(S)^{\text{fix}S}$ has a regular normal subgroup which is an elementary abelian 2-group. Since a Sylow 2-subgroup of G has order $2|S|$ it follows that $|\text{fix } S| = 2$.

LEMMA 7.5. *Let* P *be a Sylow* p-*subgroup of* $R_i \cap G_j$ *where* p *is a prime dividing* $m-2$, m, $m(m-1)$ *as the plane is of type* $(4, m)$, $(5, m)$, $(6, m)$ *respectively. Then either*

(a) $N_G(PK) \leq G_i$, *that is* $N(P^\Omega) \leq G_i^\Omega$, *and* $N(P^\Omega)$ *acts transitively on* $\text{fix}_\Omega P - \{i\}$; *or*

(b) *a Sylow* p-*subgroup of* K *has order* $|P|$, $P^\Omega = PK/K$ *is a Sylow* p-*subgroup of* G_i/K , *and* $N(P^\Omega)$ *is* 2-*transitive on* $\text{fix}_\Omega P$.

Proof. Since R_i is transitive on $\Omega - \{i\}$ it follows from [23] that $N(P^\Omega) \cap R_i^\Omega$ is transitive on fix $P - \{i\}$. By Manning [12, Theorem XIV], the number of orbits of $N(P^\Omega) = N_G(PK)/K$ in $\text{fix}_\Omega P = \text{fix}_\Omega PK$ is equal to the number of conjugacy classes of G_i/K contained in the set

$$\left\{(PK/K)^{gK} \mid (PK/K)^{gK} \leq G_i/K,\ g \in G\right\} ,$$

that is the number of conjugacy classes of G_i contained in

$$X = \left\{P^gK \mid P^gK \leq G_i,\ g \in G\right\} .$$

One such conjugacy class is $\left\{P^gK \mid g \in G_i\right\}$ and this is the set of subgroups $P'K$ where P' is a Sylow p-subgroup of R_i . If $g \in G$ is such that $P^gK \leq G_i$ then $P^g \leq R_j \cap G_i$ where $j = i^g$, and clearly X consists of the subgroups $P''K$ where P'' is a Sylow p-subgroup of $R_j \cap G_i$ for arbitrary j . Hence, since $N(P^\Omega)$ has an orbit containing fix $P - \{i\}$, part (a) is true unless X is a single conjugacy class of G_i . So suppose that all groups in X are conjugate in G_i . It follows that $PK = P'K$, where P' is a Sylow p-subgroup of $R_j \cap G_i$ for some $j \neq i$.

Thus $PK \leq R_iK \cap R_jK$. By Lemma 6.3, $R_iK \cap R_jK = K_{ij} \times K_{ji}$ where K_{ij} and K_{ji} are both isomorphic to K since $K \cap R_i = 1$, and so if a Sylow p-subgroup of K has order p^a then a Sylow p-subgroup of $R_iK \cap R_jK$ has order p^{2a} . Moreover a Sylow p-subgroup of PK has order $|P|p^a$ so that $|P| \leq p^a$. On the other hand a Sylow p-subgroup of G_{ij} has order $|P|^2$ and so $|P| \geq p^a$. Hence $|P| = p^a$ and part (b) is true by [23].

References

[1] Michael Aschbacher, "Doubly transitive groups in which the stabilizer of two points is abelian", *J. Algebra* 18 (1971), 114-136. MR43#2059.

[2] Michael Aschbacher, "On doubly transitive permutation groups of degree $n \equiv 2 \bmod 4$ ", *Illinois J. Math.* 16 (1972), 276-279. MR45#8713.

[3] Helmut Bender, "Transitive Gruppen gerader Ordnung, in denen jede Involution genau einen Punkt festlässt", *J. Algebra* 17 (1971), 527-554. MR44#5370.

[4] D. Holt, "Doubly transitive groups with a solvable one point stabiliser", preprint.

[5] D.R. Hughes, "Partial difference sets", *Amer. J. Math.* 78 (1956), 650-674. MR18,921.

[6] D.R. Hughes, "A note on some partially transitive projective planes", *Proc. Amer. Math. Soc.* 8 (1957), 978-981. MR19,876.

[7] Marshall Hall, Jr., *The Theory of Groups* (The Macmillan Co., New York, 1959). MR21#1996.

[8] N.L. Johnson, "A characterization of generalized Hall planes", *Bull. Austral. Math. Soc.* 6 (1972), 61-67. MR46#5861.

[9] N.L. Johnson and T.G. Ostrom, "Tangentially transitive planes of order 16 ", submitted.

[10] W.M. Kantor, "2-transitive designs", *Combinatorics*. Part 3: *Combinatorial Group Theory* (Proc. Advanced Study Institute, Breuekelen, 1974, 44-97. Math. Centre Tracts, 57. Math. Centrum, Amsterdam, 1974).

[11] Peter Lorimer, "A projective plane of order 16 ", *J. Combinatorial Theory Ser. A* 16 (1974), 334-347. MR49#3673.

[12] W.A. Manning, "The order of primitive groups (III)", *Trans. Amer. Math. Soc.* 19 (1918), 127-142. FdM46,179.

[13] Michael O'Nan, "A characterization of $L_n(q)$ as a permutation group", *Math. Z.* 127 (1972), 301-314. MR47#310.

[14] Michael E. O'Nan, "Normal structure of the one-point stabilizer of a doubly-transitive permutation group, I", *Trans. Amer. Math. Soc.* 214 (1975), 1-42.

[15] A. Rahilly, "Finite generalized Hall planes and their collineation groups", PhD thesis, University of Sydney, Sydney, 1973.

[16] Alan Rahilly, "The collineation of finite generalized Hall palnes", *Proc. Second Internat. Conf. Theory of Groups*, Canberra, 1973, 589-594 (Lecture Notes in Mathematics, 372. Springer-Verlag, Berlin, Heidelberg, New York, 1974). MR50#12767.

[17] Alan Rahilly, "Some translation planes with elations which are not translations", *Combinatorial Mathematics*, III.(Proc. Third Austral. Conf., Queensland, 1974, 197-209. Lecture Notes in Mathematics, 452. Springer-Verlag, Berlin, Heidelberg, New York, 1975).

[18] Alan Rahilly, "A remarkable translation plane of order 16 ", submitted.

[19] Alan Rahilly and D. Searby, "On partially transitive planes of Hughes type $(6, m)$ ", *Geometriae Dedicata* (to appear).

[20] Rimhak Ree, "A family of simple groups associated with the simple Lie algebra of type (G_2) ", *Amer. J. Math.* 83 (1961), 432-462. MR25#2123.

[21] Tosiro Tsuzuku, "On doubly transitive permutation groups of degree $1 + p + p^2$ where p is a prime number", *J. Algebra* 8 (1968), 143-147. MR36#1527.

[22] Helmut Wielandt, *Finite Permutation Groups* (translated from German by R. Bercov. Academic Press, New York, London, 1964). MR32#1252.

[23] Ernst Witt, "Die 5-fach transitiven Cruppen von Mathieu", *Abh. Math. Sem. Univ. Hamburg* 12 (1938), 256-264. FdM64,963.

Department of Mathematics,
Institute of Advanced Studies,
Australian National University,
Canberra, ACT, Australia;

Department of Mathematics,
University of Western Australia,
Nedlands, Western Australia, Australia.

Department of Mathematics,
School of Business and Social Sciences,
Gippsland Institute of Advanced Education,
Churchill, Victoria, Australia.

PROC. MINICONF. THEORY OF GROUPS 20C05
CANBERRA 1975, 112-117.

GROUPS WHOSE MODULAR GROUP RINGS HAVE SOLUBLE UNIT GROUPS

D.E. Taylor

Let F_p be the finite field of p elements, where p is a prime. The purpose of this paper is to describe those finite groups G for which the group ring F_pG has a soluble group of units. This question was first considered by Bateman [4], Motose and Tominaga [8] and then by Motose and Ninomiya [7]. The results of these authors reduce the problem to that of determining those 2-groups G for which the group $(F_3G)^*$ of units of F_3G is soluble. Writing $O_p(G)$ for the largest normal p-subgroup of G, our result is:

THEOREM. *Suppose that G is a finite group and p is a prime. Then $(F_pG)^*$ is soluble if and only if one of the following holds:*

(1) *$G/O_p(G)$ is abelian;*

(2) *$p = 2$ and $G/O_2(G) = E\langle t\rangle$, where E is an elementary abelian 3-group and t is an element of order 2 such that $txt = x^{-1}$ for all $x \in E$;*

(3) *$p = 3$ and $G/O_3(G)$ is one of the following 2-groups,*

(i) *$H\langle t\rangle$, where H is abelian of index 2 in $H\langle t\rangle$ and of exponent 4 or 8, t is an element of order 2 or 4 and $t^{-1}ht = h^3$ for all $h \in H$,*

(ii) *$H\langle t\rangle$, where H is an elementary abelian 2-group of index 2 in $H\langle t\rangle$ and t is an element of order 2 or 4,*

(iii) *the direct product of an elementary abelian 2-group with the group* $\langle u, v \mid u^4 = v^4 = [u, v, u] = [u, v, v] = [u, v]^2 = 1\rangle$ *of order* 32 .

Note that in each case $G/O_p(G)$ has an abelian subgroup of index at most 2 .

In §2 we give a complete proof of this theorem based on ideas developed in [9], thereby giving alternative proofs of some of the results of [4], [7] and [8].

1. Preliminary lemmas

Let J be the radical of F_pG . Then F_pG/J is a direct sum of rings $M\left(n_i, p^{k_i}\right)$, $1 \le i \le r$, where $M\left(n, p^k\right)$ denotes the ring of $n \times n$ matrices over the field F_{p^k} of p^k elements. It follows that $\left(F_pG\right)^*/(1+J)$ is the direct product of the corresponding general linear groups, $\mathrm{GL}\left(n_i, p^{k_i}\right)$, $1 \le i \le r$; consequently $1 + J = O_p\left(\left(F_pG\right)^*\right)$ and $(1+J) \cap G = O_p(G)$ (*cf.* [4, Theorem 4]).

LEMMA 1. *For* $n \ge 2$ *, the only groups* $\mathrm{GL}\left(n, p^k\right)$ *which are soluble are* $\mathrm{GL}(2, 2)$ *and* $\mathrm{GL}(2, 3)$.

Proof. Artin [2, Theorem 4.9].

LEMMA 2. *Suppose* $G = A \times H$ *, where* H *is non-abelian and* $O_p(G) = 1$. *If* $\left(F_pG\right)^*$ *is soluble, then either*

(i) $p = 2$ *and* $A = 1$ *, or*

(ii) $p = 3$ *and* A *is an elementary abelian 2-group.*

Proof (*cf.* [4, Lemma 3]). Suppose that F_pA has $M\left(n, p^k\right)$ as a homomorphic image. Since $\left(F_pH\right)^*$ is soluble and H is non-abelian with $O_p(H) = 1$ it follows from Lemma 1 that $p = 2$ or 3 and F_pH has $M(2, 2)$ or $M(2, 3)$ as a homomorphic image. Thus $F_pG = \left(F_pA\right) \otimes \left(F_pH\right)$ has $M\left(2n, p^k\right)$ as a homomorphic image, whence $n = k = 1$ and F_pA is a direct sum of copies of F_p .

LEMMA 3. *If* A_4 *is the alternating group on* 4 *letters, then* $\left(F_3A_4\right)^*$ *is not soluble.*

Proof. Since $\mathrm{GL}(2, 3)$ does not contain any subgroup isomorphic to A_4 , dimension considerations show that $F_3A_4/J = F_3 \oplus M(3, 3)$. Since $\mathrm{GL}(3, 3)$ is not soluble, neither is $\left(F_3A_4\right)^*$.

Let D_8 denote the dihedral group of order 8, Q_8 the quaternion group of order 8, Z_n the cyclic group of order n, and

$$S_{16} = \langle a, b \mid a^2 = b^8 = 1, ab = b^3a\rangle$$

the semidihedral group of order 16. Note that S_{16} is the Sylow 2-subgroup of $\mathrm{GL}(2, 3)$.

LEMMA 4. *If G is a subdirect product of Z_8 with either D_8 or Q_8, then $(F_3G)^*$ is not soluble.*

Proof (*cf.* [4, Theorem 12]). This lemma follows from Lemma 2 if G is the direct product. If G is not the direct product, then $|G| = 2^5$ and $Z(G) = G' \times H$, where H is cyclic of order 4 and G' has order 2. If N is a normal subgroup of G not contained in $Z(G)$, then N contains G'. It follows that the kernels of the representations of G into $M(2, 3)$ lie in $Z(G)$ and hence F_3G has just two direct summands isomorphic to $M(2, 3)$. By calculating the dimension we see that F_3G must have $M(2, 9)$ as a direct summand and hence $(F_3G)^*$ cannot be soluble.

LEMMA 5. *Let R and S be rings with unit groups R^* and S^* respectively and suppose that $\phi : R \to S$ is a surjective homomorphism. If the kernel ϕ is nilpotent or if $R/\mathrm{rad}\, R$ is Artinian, then $\phi(R^*) = S^*$.*

Proof. Bass [3, p. 87].

In the proof of the theorem, this lemma allows us to apply induction to factor groups of G.

2. Proof of the theorem

If $G/O_p(G)$ is abelian, $(F_pG)^*$ is certainly soluble. Suppose $G = E\langle t\rangle$, where E is an elementary abelian 3-group of order 3^n and t is an element of order 2 which inverts E. Each of the $\frac{1}{2}(3^n-1)$ subgroups of index 3 in E is normal in G and thus each occurs as the kernel of a homomorphism of G onto $\mathrm{GL}(2, 2)$. This provides $\frac{1}{2}(3^n-1)$ direct summands of F_2G/J isomorphic to $M(2, 2)$. By calculating the dimension we see that the only other direct summand is F_2 and hence $(F_2G)^*$ is soluble. The class of 2-groups listed under (3) of the theorem is easily seen to be closed under the formation of non-abelian homomorphic images while the only such groups with cyclic centre are D_8, Q_8 and S_{16}. It follows that only $M(2, 3)$ occurs as a matrix ring direct summand of F_3G and hence

$(F_3G)^*$ is soluble.

Now suppose that $(F_pG)^*$ is soluble and $G/O_p(G)$ is non-abelian. From §1 we deduce that $p = 2$ or 3 and $(F_pG)/J$ is a direct sum of fields F_{p^k} and copies of $M(2, p)$. Moreover, in proving the theorem we may assume that $O_p(G) = 1$. If $p = 2$, G is a subdirect product of groups H and K where H is an abelian group of odd order and K is a subdirect product of copies of $\mathrm{GL}(2, 2) \simeq S_3$. Since K can have no non-trivial factor group which is abelian of odd order we have $G = H \times K$ and, from Lemma 2, $H = 1$. If t is an element of order 2 in G, then $C_{O_3(G)}(t)$ is a direct factor of $O_3(G)\langle t\rangle$ so by Lemma 2, $C_{O_3(G)}(t) = 1$ and hence t inverts $O_3(G)$. If s is an element of order 2 which commutes with t then st commutes with every element of $O_3(G)$ and therefore $st = 1$; that is, $s = t$. It follows that $G = O_3(G)\langle t\rangle$ and we have (2) of the theorem.

This leaves us with the case $p = 3$. Since $\mathrm{SL}(2, 3)$ has A_4 as a homomorphic image it follows from Lemma 3 and the assumption $O_3(G) = 1$ that the image of G in each $M(2, 3)$ summand of F_3G/J is D_8, Q_8 or S_{16}. From Lemmas 2 and 4, G is a subdirect product of copies of Z_2, Z_4, D_8, Q_8 and S_{16}. In particular, G is a 2-group. We complete the proof by induction on the order of G.

Suppose at first that there is an element x of order 4 such that $\langle x\rangle$ is normal in G. We shall show by induction that G is one of the groups listed in (3) (i) of the theorem. Choose a normal subgroup N such that $N \cap \langle x\rangle = 1$ and G/N is isomorphic to Z_4, D_8, Q_8 or S_{16}. If $G/N \simeq Z_4$, then $G = \langle x\rangle \times N$, in contradiction to Lemma 2. It follows that $Z(G)$ is elementary abelian and by induction $C_G(x)$ is abelian and has index 2 in G. If $t \in G\text{-}H$, then $t^2 \in Z(G)$ so that the order of t is 2 or 4. If $N = 1$, then G is D_8, Q_8 or S_{16} and each of these possibilities occurs in (3) (i) of the theorem. Suppose $N \neq 1$ and choose $z \in N \cap Z(G)$, $z \neq 1$. By induction applied to $G/\langle z\rangle$ and $G/\langle zx^2\rangle$ we find that for all $h \in H$, $t^{-1}hth^{-3} \in \langle z\rangle \cap \langle zx^2\rangle = 1$ and again G occurs in (3) (i) of the theorem. The square of any element in D_8, Q_8 or S_{16} generates a normal subgroup so if $y \in G$ has order 8, $\langle y^2\rangle$ is a normal subgroup of order 4.

From now on we assume G has no cyclic normal subgroup of order 4. This means that G is a subdirect product of copies of Z_2, Z_4, D_8 and Q_8 ; in particular, both $Z(G)$ and $G/Z(G)$ are elementary abelian. If $u \notin Z(G)$, then by induction we have $C_G(u)' \subseteq \langle z \rangle$ or $[G, u] \subseteq \langle z \rangle$ for each element $1 \neq z \in Z(G)$. If $C_G(u)$ is not abelian, then $|Z(G)| = 2$ and hence $G = D_8$ or Q_8 , contrary to assumption. It follows that $C_G(u)$ is abelian for all $u \notin Z(G)$.

Suppose that K is a normal subgroup of G such that G/K is non-abelian. If K is also non-abelian, we can find N such that $G/N \simeq D_8$ or Q_8 and KN/N is non-abelian. It follows that $KN = G$ and consequently $G/K \cap N \simeq G/K \times G/N$, contrary to Lemma 2.

Suppose now that no maximal subgroup of G is abelian. We shall prove that $|G/Z(G)| = 8$. (This follows from a result of Amitsur [1], but in our context it is easy to give a direct proof.) If M is a maximal subgroup of G , then by induction M has an abelian subgroup A of index 2 and it follows that $A \supseteq Z(G)$. Moreover, since $A \neq Z(G)$ we can find a normal subgroup N such that $G/N \simeq D_8$ or Q_8 and $|AN/N| = 4$. Since N is abelian we see that $A \cap N \subseteq Z(G)$ and since $G' \subseteq Z(G)$ we have $Z(G) \nsubseteq N$. Thus $|G : A \cap N| = 16$ and we have $|G : Z(G)| = 8$. Now each element $u \notin Z(G)$ has 4 conjugates, so G has $\frac{1}{4}(|G|-|Z(G)|) + |Z(G)|$ conjugacy classes. The number of characters of degree 1 is $|G/G'|$ and the remaining irreducible characters of G all have degree 2 , so G has $\frac{1}{4}(|G|-|G/G'|) + |G/G'|$ irreducible characters. Equating the number of characters with the number of classes we deduce that $|G'| = 8$. (This argument comes from [6, p. 454].) By factoring a complement to G' in $Z(G)$ we obtain a group with $G' = Z(G)$ and hence $|G| = 2^6$. In the notation of Hall and Senior [5], G is in the family Γ_9 and therefore has a non-abelian factor group with an element of order 4 in the centre. This contradiction shows that some maximal subgroup A of G is abelian. If a maximal subgroup of G is elementary abelian we have a group of (3) (ii) of the theorem. We assume this is not the case and choose an element $u \in A$ of order 4 . If $t \notin A$, then $[u, t] \neq u^2$ so the image of u in $G/\langle [u, t] \rangle$ is a central element of order 4 . It follows that $G/\langle [u, t] \rangle$ is abelian, and hence $|G'| = 2$. Now for all $x \notin Z(G)$, $C_G(x)$ is abelian and has index 2 in G , therefore every element of $G - Z(G)$ has order 4 . Choose $u \notin Z(G)$ and $v \notin C_G(u)$. Then $Z(G) = C_G(u) \cap C_G(v)$ has index 4 in G . If either $u^2 = [u, v]$, $v^2 = [u, v]$ or $u^2 = v^2$, then either $\langle u \rangle$, $\langle v \rangle$ or $\langle uv \rangle$ is normal in G , respectively, contrary to assumption. It follows that $\langle u^2, v^2, [u, v] \rangle$ is a group of order 8 and we have (3) (iii) of the theorem.

References

[1] S.A. Amitsur, "Groups with representations of bounded degree II", *Illinois J. Math.* 5 (1961), 198-205. MR23#A225.

[2] E. Artin, *Geometric Algebra* (Interscience, New York, London, 1957). MR18,553.

[3] Hyman Bass, *Algebraic K-Theory* (Benjamin, New York, Amsterdam, 1968). MR40#2736.

[4] J.M. Bateman, "On the solvability of unit groups of group algebras", *Trans. Amer. Math. Soc.* 157 (1971), 73-86. MR43#2118.

[5] Marshall Hall, Jr. and James K. Senior, *The Groups of Order* 2^n $(n \leq 6)$ (Macmillan, New York; Collier-Macmillan, London, 1964). MR29#5889.

[6] I.M. Isaacs and D.S. Passman, "Groups whose irreducible representations have degrees dividing p^e ", *Illinois J. Math.* 8 (1964), 446-457. MR29#3552.

[7] Kaoru Motose and Yasushi Ninomiya, "On the solvability of unit groups of group rings", *Math. J. Okayama Univ.* 15 (1972), 209-214. MR48#397.

[8] Kaoru Motose and Hisao Tominaga, "Group rings with solvable unit groups", *Math. J. Okayama Univ.* 15 (1971/72), 37-40. MR46#5423.

[9] K.R. Pearson and D.E. Taylor, "Groups subnormal in the units of their modular group rings", *Proc. London Math. Soc.* (3) (to appear).

Department of Pure Mathematics,
University of Sydney,
Sydney,
New South Wales, Australia.

PROC. MINICONF. THEORY OF GROUPS
CANBERRA 1975, 118-125. 20D10

COMPUTING SOLUBLE GROUPS

J.W. Wamsley

An algorithm is given for computing soluble groups which is an extension of the nilpotent quotient algorithm. The method is based on the Reidemeister-Schreier method of presenting subgroups. Given $G/G^{(k)}$, where $G^{(k)}$ is the k-th term in the derived series we construct $G^{(k)}/G^{(k+1)}$ and hence extend $G/G^{(k)}$ to $G/G^{(k+1)}$. The problem of many generators is partially solved by introducing the notion of a module presentation.

1. The Reidemeister-Schreier method

We give an outline of this method, as we will use it. A full description may be found in [1].

Let $G = \{x_1, \ldots, x_n \mid R_1, \ldots, R_m\} = F/R$, where F is free on $x_1, \ldots, x_n$ and G/N a factor group with presentation

$$G/N = \{x_1, \ldots, x_n \mid R_1, \ldots, R_m, s_1, \ldots, s_t\} = F/S .$$

Let $\{g_i\}$ be a two-sided Schreier system for S in F and we will write (f) to represent the representative of f in the Schreier system. Then S, and therefore N, is generated by the set $\left\{\left(x_ig_j\right)^{-1}x_ig_j \neq 1\right\}$. The number of these generators is $1+(n-1)|G/N|$ and S is free of this rank. Also we can calculate the action of F under conjugation on S by simply writing $x_k^{-1}\left(x_ig_j\right)^{-1}x_ig_jx_k$ as a word in the free generators of S.

Suppose we write $t_{ij} = \left(x_ig_j\right)^{-1}x_ig_j$, then $x_k^{-1}t_{ij}x_k = \omega_{ijk}\left(t_{\alpha\beta}\right)$ and we have

$$F = \left\{x_1, \ldots, x_n, t_{ij} \mid t_{ij} = (x_i g_j)^{-1} x_i g_j,\ x_k^{-1} t_{ij} x_k = \omega_{ijk}(t_{\alpha\beta})\right\} .$$

We have that N is generated by the set $\{t_{ij}\}$ and we know how G/N is extended by N. A set of defining relations for N is given by the set $\left\{g_i^{-1} R_j g_i\right\}$ where each is written in the t_{ij}.

We give an example.

Let

$$G = \{a, b \mid a^2 = b^3 = (ab)^2 = 1\} = F/R$$

and

$$G/N = \{a, b \mid a^2 = b = 1\} = F/S .$$

S has free generators $\{a^2, b, a^{-1}ba\}$,

$$F = \{a, b, r, s, t \mid a^2 = r, b = s, a^{-1}ba = t, a^{-1}ra = r, a^{-1}sa = t, a^{-1}ta = r^{-1}sr,$$
$$b^{-1}rb = s^{-1}rs, b^{-1}sb = s, b^{-1}tb = s^{-1}ts\}$$

and

$$G = \{a, b, r, s, t \mid a^2 = r, b = s, a^{-1}ba = t, a^{-1}ra = r, a^{-1}sa = t, a^{-1}ta = r^{-1}sr,$$
$$b^{-1}rb = s^{-1}rs, b^{-1}sb = s, b^{-1}tb = s^{-1}ts, a^2 = b^3 = (ab)^2 = 1\} .$$

Now $a^2 = r$, $b^3 = s^3$, $(ab)^2 = rts$, therefore in G, we have $r = 1$, $s^3 = 1$ and $ts = 1$. Conjugation of r, s^3 and ts in F by a and b yield no new relation, so we have

$$N = \{r, s, t \mid r = 1, s^3 = 1, ts = 1\}$$

and

$$G = \{a, b, s \mid a^2 = 1, b = s, a^{-1}sa = s^2, s^3 = 1\} .$$

In general the presentation for N will be so messy that we won't know what N is. If, however, G is soluble then we may assume N is abelian and since abelian groups are "easy" to handle we may for example go down the derived series for G and hence obtain G. If $G/G^{(k)} = F/S$ where $|G/G^{(k)}| = g$ and F is free on d generators then S/S' will be free abelian on $1 + (d-1)g$ generators. If this number, $(1+(d-1)g)$, is reasonably small then we may without difficulty write a computer programme to calculate $G/G^{(k+1)}$. However, in general, this number will be too large. We therefore look at module presentations.

2. Module presentations

Let G be a group and M a finitely generated G-module then a presentation for M is

$$M = \{m_1, \ldots, m_r \mid t_1, \ldots, t_s\} = F/T$$

where F is the free G-module generated by $m_1, \ldots, m_r$ and T is the submodule of F generated by $t_1, \ldots, t_s$.

For example in the previous presentation for S_3 we have

$$S/S' = \{r, s, t \mid r \circ a = r, s \circ a = t, t \circ a = s\} = F/T$$

where $F = ZG \oplus ZG \oplus ZG$, $G = \{a \mid a^2 = 1\}$, and T is the submodule of F generated by

$$(1-a, 0, 0), (0, -a, 1) \text{ and } (0, 1, -a) ,$$

which we will denote in matrix form as

$$\begin{pmatrix} 1-a & 0 & 0 \\ 0 & -a & 1 \\ 0 & 1 & -a \end{pmatrix} .$$

This gives a presentation for N/N' to be that of S/S' together with relations $r = 0$, $3s = 0$ and $s + t = 0$ or in matrix form

$$\begin{pmatrix} 1-a & 0 & 0 \\ 0 & -a & 1 \\ 0 & 1 & -a \\ 1 & 0 & 0 \\ 0 & 3 & 0 \\ 0 & 1 & 1 \end{pmatrix} .$$

We may carry out the normal Tietze transformations and thereby "diagonalise" the matrix. In this case it becomes

$$\begin{pmatrix} 1 & 0 & 0 \\ 0 & 1 & 1 \\ 0 & 0 & a+1 \\ 0 & 0 & 3 \end{pmatrix} .$$

This gives that N/N' is cyclic of order 3 , generated by t with $a^{-1}ta = t^2$.

3. Soluble groups

Let G be a soluble group then G has a presentation

$$G = \left\{a_1, \ldots, a_n \mid a_i^{\rho_i} = a_{i+1}^{\alpha_{i+1}} \ldots a_n^{\alpha_n},\ 1 \le i \le n,\right.$$

$$\left.[a_j, a_i] = a_{i+1}^{\beta_{i+1}} \ldots a_n^{\beta_n},\ 1 \le i < j \le n\right\} = F/R\ .$$

The equations

$$a_i^{\rho_i} = a_{i+1}^{\alpha_{i+1}} \ldots a_n^{\alpha_n} r_i\ ,$$

$$[a_j, a_i] = a_{i+1}^{\beta_{i+1}} \ldots a_n^{\beta_n} s_{ji}$$

will be called the extended relations.

Collection of a word in G will involve substituting $a_{i+1}^{\alpha_{i+1}} \ldots a_n^{\alpha_n}$ for $a_i^{\rho_i}$ or $a_i a_j a_{i+1}^{\beta_{i+1}} \ldots a_n^{\beta_n}$ for $a_j a_i$. A word is said to be collected when it is in the form

$$a_1^{\gamma_1} \ldots a_n^{\gamma_n}\ ,\quad 0 \le \gamma_i < \rho_i\ .$$

A presentation for G is said to be consistent if the collected form of a word is unique.

We have

THEOREM. *Let G be a finite soluble group with consistent presentation as above, then R/R' is a G-module with presentation given by generators $\{r_i, s_{kj}\}$ and relations given by collecting the consistency equations*

(1) $(a_k a_j) a_i = a_k (a_j a_i)\ ,\quad 1 \le i < j < k \le n\ ,$

(2) $(a_k a_j) a_j^{\rho_j - 1} = a_k \left(a_j^{\rho_j}\right)\ ,\quad 1 \le j < k \le n$ *and*

(3) $\left(a_k^{\rho_k}\right) a_j = a_k^{\rho_k - 1} (a_k a_j)\ ,\quad 1 \le j \le k \le n\ ,$

using the extended relations, where the bracketted part is collected first and then collection is carried out in any convenient manner. //

THEOREM. *Let G be a group with presentation $G = \{a_1, \ldots, a_r \mid R_1, \ldots, R_s\}$, and suppose we have a consistent presentation for $G/G^{(k)} = F/R$, then the module $G^{(k)}/G^{(k+1)}$ has a presentation given by a presentation for R/R' with relations added, given by collecting $R_1, \ldots, R_s$ using the extended relations and setting definitions equal to 1.*

4. An example

Let

$$G = \{a, b, c \mid [b, a] = b^3, [c, b] = c^3, [c, a] = a^3\} .$$

It is known that G'' is abelian.

$$G/G' = \{a, b, c \mid a^3 = 1, b^3 = 1, c^3 = 1, [b, a] = 1, [c, b] = 1, [c, a] = 1\}$$
$$= F/R$$

then R/R' is a free abelian group of rank 55 .

However, R/R' has presentation given by the matrix where the columns correspond to $a^3 = \alpha$, $b^3 = \beta$, $c^3 = \gamma$, $[b, a] = \delta$, $[c, b] = \lambda$ and $[c, a] = \mu$ respectively.

$$\begin{pmatrix} a-1 & 0 & 0 & 0 & 0 & 0 \\ 0 & 1-a & 0 & 1+b+b^2 & 0 & 0 \\ b-1 & 0 & 0 & 1+a+a^2 & 0 & 0 \\ 0 & 0 & 1-a & 0 & 0 & 1+c+c^2 \\ c-1 & 0 & 0 & 0 & 0 & 1+a+a^2 \\ 0 & b-1 & 0 & 0 & 0 & 0 \\ 0 & 0 & 1-b & 0 & 1+c+c^2 & 0 \\ 0 & c-1 & 0 & 0 & 1+b+b^2 & 0 \\ 0 & 0 & c-1 & 0 & 0 & 0 \\ 0 & 0 & 0 & c-1 & a-1 & 1-b \end{pmatrix} \begin{array}{l} (a^3)a = a(a^3) \\ (b^3)a = b^2(ba) \\ b(a^3) = (ba)a^2 \\ (c^3)a = c^2(ca) \\ c(a^3) = (ca)a^2 \\ (b^3)b = b(b^3) \\ (c^3)b = c^2(cb) \\ c(b^3) = (cb)b^2 \\ (c^3)c = c(c^3) \\ (cb)a = c(ba) \ . \end{array}$$

G'/G'' has the same presentation with the added rows corresponding to the defining relation $[b, a] = b^3$, $[c, b] = c^3$ and $[c, a] = a^3$ respectively; that is,

$$\begin{pmatrix} 0 & 1 & 0 & -1 & 0 & 0 \\ 0 & 0 & 1 & 0 & -1 & 0 \\ 1 & 0 & 0 & 0 & 0 & -1 \end{pmatrix} .$$

Row operations yield that G'/G'' is generated by α, β and γ (with $\delta = \beta$, $\lambda = \gamma$ and $\mu = \alpha$), with defining relations given by the rows

$$\begin{pmatrix} a-4 & 0 & 0 \\ b-1 & 0 & 0 \\ c-1 & 21 & 0 \\ 21 & -21 & 3 \\ 0 & a-1 & -3 \\ 0 & b-4 & 0 \\ 0 & c-1 & 0 \\ 0 & 21 & b-4 \\ 0 & 63 & 0 \\ 0 & 0 & a-1 \\ 0 & 0 & c+2 \\ 0 & 0 & 9 \end{pmatrix};$$

that is,

$$G/G' = \{a, b, c, x, y, z \mid a^3 = z, b^3 = x, c^3 = y, [b, a] = x, [c, b] = y, [c, a] = z, [x, a] = x^3, [x, b] = 1, [x, c] = y^{42}, x^{21} = y^{21}z^6, [y, a] = z^3, [y, b] = y^3, [y, c] = 1, y^{63} = 1, [z, a] = 1, [z, b] = y^{42}z^3, [z, c] = z^6, z^9 = 1, [y, x] = 1, [z, y] = 1, [z, x] = 1\}$$

$= F/R$ where R/R' is a free abelian group of rank 1,607,446 .

However, carrying out consistency equations in the right order yields:

Letting $x^{21} = y^{21}z^6\alpha$, $y^{63} = \beta$, $z^9 = \gamma$ we have

$$G = \{a, b, c, x, y, z, \alpha, \beta, \gamma\}$$

with defining relations

(1)	$[b, a] = x$	definition
(2)	$[c, b] = y$	definition
(3)	$[c, a] = z$	definition
(4)	$a^3 = z$	relation
(5)	$b^3 = x$	relation
(6)	$c^3 = y$	relation
(7)	$[x, b] = 1$	$(b^3)b = b(b^3)$
(8)	$[x, a] = x^3$	$(b^3)a = b^2(ba)$
(9)	$[y, c] = 1$	$(c^3)c = c(c^3)$
(10)	$[y, b] = y^3$	$(c^3)b = c^2(cb)$
(11)	$[x, c] = y^{42}\beta^{-1}$	$c(b^3) = (cb)b^2$
(12)	$[z, a] = 1$	$(a^3)a = a(a^3)$

(13) $[z, c] = z^{6}.\gamma^{-1}$ $\qquad c(a^{3}) = (ca)a^{2}$

(14) $[y, a] = z^{3}$ $\qquad (c^{3})a = c^{2}(ca)$

(15) $[z, b] = \alpha^{-1}z^{-6}y^{-21}$ $\qquad b(a^{3}) = (ba)a^{2}$

(16) $[z, y] = \gamma^{-1}$ $\qquad y(a^{3}) = (ya)a^{2}$

(17) $[\beta, c] = 1$ $\qquad (y^{63})c = y^{62}(yc)$

(18) $[\beta, y] = 1$ $\qquad \beta(c^{3}) = (\beta c)c^{2}$

(19) $[y, x] = \beta$ $\qquad x(c^{3}) = (xc)c^{2}$

(20) $[z, x] = \alpha^{-1}.z^{-1}y^{-21}\alpha^{-1}y^{21}z^{6}.z^{-12}y^{-42}\alpha^{-1}y^{42}z^{12}.$

$$\beta^{-1}.\gamma^{\frac{-6+6(8)^{21}-6(8)^{42}}{9}} = x^{-63} \qquad x(a^{3}) = (xa)a^{2}$$

Note that, although we have effectively reduced to 1345 free generators, the numbers appearing are becoming enormous and we must at some stage put a bound on these. Since it is known that G'' is of exponent 3 , I will now use this fact.

(21) $[\gamma, z] = 1$ $\qquad z(z^{9}) = (z^{9})z$

(22) $[\alpha, a] = \alpha\beta\gamma.z^{-6}y^{-21}\alpha y^{21}z^{6}.z^{-12}y^{-42}\alpha y^{42}z^{12}$ $\qquad (x^{21})a = x^{20}(xa)$

(23) $[\beta, a] = 1$ $\qquad (y^{63})a = y^{62}(ya)$

(24) $[\gamma, a] = 1$ $\qquad (z^{9})a = z^{8}(za)$

(25) $[\beta, b] = 1$ $\qquad (y^{63})b = y^{62}(yb)$

(26) $[\alpha, z] = 1$ $\qquad z(x^{21}) = (zx)x^{20}$

(27) $[\alpha, y] = 1$ $\qquad y(x^{21}) = (yx)x^{20}$

(28) $[z, x] = \beta^{-1}\gamma$ $\qquad$ (20)

(29) $[\alpha, a] = \beta\gamma$ $\qquad$ (22)

(30) $[\alpha, b] = \beta\gamma$ $\qquad (x^{21})b = x^{20}(xb)$

(31) $[\gamma, b] = 1$ $\qquad (z^{9})b = z^{8}(zb)$

(32) $[\gamma, c] = 1$ $\qquad (z^{9})c = z^{8}(zc)$

(33) $[\alpha, c] = \gamma^{2}\beta^{2}$ $\qquad (x^{21})c = x^{20}(xc)$

(34) $[\beta, z] = 1$ $\qquad z(y^{63}) = (zy)y^{62}$

(35) $[\alpha, x] = 1$ $\qquad (x^{21})x = x(x^{21})$

(36) $[\beta, x] = 1$ $\qquad (y^{63})x = y^{62}(yx)$

(37) $[\gamma, x] = 1$ $\qquad (z^{9})x = z^{8}(zx)$

(38) $[\gamma, y] = 1$ $\qquad (z^{9})y = z^{8}(zy)$

(39) $\gamma = \alpha$ $\qquad (cb)a = c(ba)$

(40) $\beta\gamma = 1$ $\qquad$ (24 & 29).

Hence we have that G'' is spanned by α with a, b and c acting trivially on it and α of order 3.

Nothing is gained from the other consistency equations and we have

$$
\begin{aligned}
G = \{a, b, c, x, y, z, \alpha \mid\ & [b, a] = x,\ [c, b] = y,\ [c, a] = z \\
& a^3 = z,\ b^3 = x,\ c^3 = y,\ x^{21} = y^{21} z^{\alpha} \alpha,\ y^{63} = \alpha^2, \\
& z^9 = \alpha,\ [y, x] = \alpha^2,\ [z, y] = \alpha^2,\ [z, x] = \alpha^2, \\
& [x, a] = x^3,\ [x, b] = 1,\ [x, c] = y^{42}\alpha,\ [y, a] = z^3,\ [y, b] = y^3, \\
& [y, c] = 1,\ [z, a] = 1,\ [z, b] = z^3 y^{42} \alpha^2,\ [z, c] = z^6 \alpha^2,\ \alpha^3 = 1,\ \alpha \text{ central}\};
\end{aligned}
$$

which leads to thc following interesting question. Suppose we let $D^1(G) = G$, $D^{i+1}(G) = [D^i(G), D^i(G)]$. Let G be a 3 generator 3 relation finite group, is it true that $D^3(G)$ is a 3-group?

Reference

[1] Wilhelm Magnus, Abraham Karrass, Donald Solitar, *Combinatorial Group Theory: Presentations of groups in terms of generators and relations* (Pure and Applied Mathematics, 13. Interscience [John Wiley & Sons], New York, London, Sydney, 1966). MR34#7617.

School of Mathematics,
Flinders University,
Bedford Park,
South Australia, Australia.

PROC. MINICONF. THEORY OF GROUPS
CANBERRA 1975, 126-129. 20E99

A NOTE ON THE TODD-COXETER ALGORITHM

J.N. Ward

The Todd-Coxeter algorithm is described in [1], §§2.1 and 2.2. The description given there asks that before defining a new coset, each coset must appear "in every essentially different place" in the relator tables. However in each of the examples given it is sufficient to ensure that each coset appears at the start and end of some row under each relator. The question arises as to whether this is sufficient in all cases. The purpose of this note is to show, by means of some examples, that it is not.

The first example is a representation of the infinite cyclic group:

$$G = \langle X, Y \mid YXYXY^{-1} = 1 \rangle .$$

For the subgroup H whose index we wish to compute we take $\langle X \rangle$.

Since $YXYXY^{-1} = 1$ we get $XYX = 1$ or $Y = X^{-2}$. Thus $|G : H| = 1$.

If we use the method indicated above we obtain the following coset multiplication and relator table:

Multiplication table

	X	Y	X^{-1}	Y^{-1}
1	1	2	1	
2	3	5	4	1
3		4	2	
4	2	6	5	3
5	4	7	6	2
6	5	8	7	4
$n \geq 7$	$n-1$	$n+2$	$n+1$	$n-2$

Relator table

	Y	X	Y	X	Y^{-1}
1	2	3	4	2	1
2	5	4	6	5	2
3	4	2	5	4	3
4	6	5	7	6	4
5	7	6	8	7	5
6	8	7	9	8	6
$n \geq 7$	$n+2$	$n+1$	$n+3$	$n+2$	n

The reason that the algorithm does not terminate in this example is that we are essentially enumerating the cosets of the identity in the infinite cyclic group G. The extra information, $1X = 1$, which we obtain in the multiplication table because we are enumerating the cosets of $\langle X \rangle$ is never used in the relator table.

The second example is a little harder to see but in it we show that the above situation is not the only one in which the procedure can fail. In fact we give a presentation of the cyclic group of order 3 such that the procedure does not terminate when we attempt to enumerate the cosets of the identity.

Let $G = \langle X, Y \mid X^3YXYX^{-3} = Y^3XYXY^{-3} = 1 \rangle$. Since $X^3YXYX^{-3} = 1$ yields $X = Y^{-2}$ and $Y^3XYXY^{-3} = 1$ yields $Y = X^{-2}$ we see that G is cyclic of order 3.

To show that the procedure will not terminate if we attempt to enumerate the cosets of the identity we will define a set S of objects with the following properties:

(i) if $E \in S$ then EX and EY are also elements of S;

(ii) if $E \in S$ and $E = FX$ (FY) for some $F \in S$ then F is uniquely determined;

(iii) for all $E \in S$ we have $EX^3YXY = EX^3$ and $EY^3XYX = EY^3$;

(iv) there exists an element $0 \in S$ such that if $E \in S$ then $E = 0W$ where W is some sequence of the letters X and Y;

(v) $|S|$ is infinite.

It is not difficult to prove that the existence of such a set S will imply that the procedure does not terminate since every coset we define in the procedure can be mapped onto one of the elements of S and given any element $E \in S$ we will ultimately define a coset which maps onto E.

We now construct such a set S with the properties listed. Let $I = \{0, 1, 2, \ldots\}$ and set

$$A = \{\mathrm{a} \mid \mathrm{a} = (a_1, a_2, \ldots) \text{ where } a_i \in I \text{ for each } i,\ a_k = 0 \text{ for some } k \geq 2 \text{ and if } a_j = 0 \text{ for } j \geq 2 \text{ then } a_i = 0 \text{ for all } i \geq j\}.$$

We define a relation $\sim$ on A as follows.

Assume $\mathrm{a} = (a_1, a_2, \ldots)$ and $\mathrm{b} = (b_1, b_2, \ldots)$ are elements of A. We say $\mathrm{a} \sim \mathrm{b}$ if for some integer r,

$$a_r \geq 3,\quad 1 = a_{r+1} = a_{r+2} \leq a_{r+3}$$

and either

$$\left.\begin{array}{lll} b_s = a_s & & s \le r \\ b_{r+1} = a_{r+3} - 1 & & \\ b_s = a_{s+2} & & s \ge r+2 \end{array}\right\} \text{ if } a_{r+3} > 1$$

or

$$\left.\begin{array}{lll} b_s = a_s & & s \le r-1 \\ b_r = a_r + a_{r+4} & & \\ b_s = a_{s+4} & & s \ge r+1 \end{array}\right\} \text{ if } a_{r+3} = 1 .$$

Notice that if $\mathrm{a}, \mathrm{b} \in A$ and $\mathrm{a} \sim \mathrm{b}$ then for some r, $a_r \ge 3$ and $b_r \ge 3$. This means, for example, that the sequence α_n which has 1 in the first n positions and zeros thereafter does not satisfy $\alpha_n \sim \mathrm{a}$ or $\mathrm{a} \sim \alpha_n$ for any sequence $\mathrm{a} \in A$.

Let $\mathrm{a} \in A$ and assume n is the largest integer such that $a_n \ne 0$. We define

$$\begin{aligned} \mathrm{a}X &= (a_1, \ldots, a_n, 1, 0, 0, \ldots) \text{ if } n \text{ is even} \\ &= (a_1, \ldots, a_{n-1}, a_n+1, 0, 0, \ldots) \text{ if } n \text{ is odd} \end{aligned}$$

and

$$\begin{aligned} \mathrm{a}Y &= (a_1, \ldots, a_{n-1}, a_n+1, 0, 0, \ldots) \text{ if } n \text{ is even} \\ &= (a_1, \ldots, a_n, 1, 0, 0, \ldots) \text{ if } n \text{ is odd.} \end{aligned}$$

If $\mathrm{a} = (0, 0, \ldots)$ we define

$$\mathrm{a}X = (1, 0, 0, \ldots)$$

and

$$\mathrm{a}Y = (0, 1, 0, \ldots) .$$

It is easily checked that if $\mathrm{a} \sim \mathrm{b}$ then $\mathrm{a}X \sim \mathrm{b}X$ and $\mathrm{a}Y \sim \mathrm{b}Y$.

We now define an equivalence relation $\simeq$ on A by letting $\mathrm{a} \simeq \mathrm{b}$ if either $\mathrm{a} = \mathrm{b}$ or there exists a sequence $\mathrm{a}_0, \mathrm{a}_1, \ldots, \mathrm{a}_n$ of elements of A such that $\mathrm{a}_0 = \mathrm{a}$, $\mathrm{a}_n = \mathrm{b}$ and for each i, $1 \le i \le n$, we have either $\mathrm{a}_i \sim \mathrm{a}_{i-1}$ or $\mathrm{a}_{i-1} \sim \mathrm{a}_i$.

Clearly if $\mathrm{a} \simeq \mathrm{b}$ then $\mathrm{a}X \simeq \mathrm{b}X$ and $\mathrm{a}Y \simeq \mathrm{b}Y$. Conversely if $\mathrm{a}X \simeq \mathrm{b}X$ (or $\mathrm{a}Y \simeq \mathrm{b}Y$) then $\mathrm{a} \simeq \mathrm{b}$.

Let, for each $\mathrm{a} \in A$, $[\mathrm{a}]$ denote the equivalence class of A containing a with the equivalence relation $\simeq$.

Now let S denote the set of equivalence classes of A under the relation $\simeq$.

If $E \in S$ and $\mathbf{a} \in E$ define

$$EX = [\mathbf{a}X]$$

and

$$EY = [\mathbf{a}Y] .$$

S is easily seen to satisfy properties (i) and (ii) and property (iii) follows from the definition of $\sim$. For property (iv) take $0 = [(0, 0, 0, \ldots)]$. Then if $E = [\mathbf{a}]$ where $\mathbf{a} = (a_1, a_2, \ldots, a_{2n}, 0, 0, 0, \ldots)$ we have

$$E = 0X^{a_1}Y^{a_2} \ldots X^{a_{2n-1}}Y^{a_{2n}} .$$

Finally S is infinite since for each integer n the equivalence class $[\boldsymbol{\alpha}_n]$ contains just the single element $\boldsymbol{\alpha}_n$ and the $\boldsymbol{\alpha}_n$ are all distinct. Thus S satisfies all the five properties and the proof of our assertion is complete.

Reference

[1] H.S.M. Coxeter and W.O.J. Moser, *Generators and Relations for Discrete Groups*, Second Edition (Ergebnisse der Mathematik und ihrer Grenzgebiete, 14. Springer-Verlag, Berlin, Göttingen, Heidelberg, New York, 1965).

Department of Pure Mathematics,
University of Sydney,
Sydney,
New South Wales, Australia.

PROC. MINICONF. THEORY OF GROUPS
CANBERRA 1975, 130-142.

20E15
(20E25)

LOCALLY SOLUBLE GROUPS SATISFYING THE MINIMAL CONDITION FOR NORMAL SUBGROUPS

John S. Wilson

My objectives in this talk today are to outline the current state of knowledge about locally soluble groups satisfying Min-n, the minimal condition for normal subgroups, to draw attention to some open questions concerning these groups, and to speculate about future developments. For the most part I will be discussing results proved during the last five or six years, and I will be saying very little about locally soluble groups satisfying the minimal condition for subgroups, apart from reminding you of their basic properties.

(A) (Černikov, 1940, 1949). *The following properties of a group G are equivalent:*

(a) G is a locally soluble group satisfying the minimal condition for subgroups,

(b) G has a finite series, each factor in which is either a finite cyclic group or a quasicyclic group, and

(c) G is a finite soluble extension of a direct product of finitely many quasicyclic groups.

In particular, locally soluble groups satisfying the minimal condition for subgroups are soluble.

By a *quasicyclic group*, I mean a group of type p^∞ for some prime p.

The groups satisfying the conditions of (A) are called *soluble Černikov groups*. There is a large and continuing literature concerned with the determination of conditions on locally soluble groups satisfying Min-n which imply the conditions of (A); among such conditions we may mention (a) the property of being locally nilpotent (Čarin, 1953), more generally, (b) the property of being locally polycyclic and having the ranks of chief factors of finite subgroups bounded (McLain, 1959), and, by way of contrast, (c) the property of being a group of automorphisms of a soluble group of finite rank (an easy extension of a theorem of Murač, 1973). An example of a (locally) soluble group satisfying Min-n which is not a soluble Černikov group was first found by Čarin, and I would like to describe this in detail.

Let p and q be distinct primes, and let F be the subfield of the algebraic closure of the field of p elements generated by the set Q of all qth power roots of unity; then Q is a quasicyclic group, and Q acts in a natural way by multiplication on the additive group A of F. A routine argument shows that Q acts irreducibly on A, and that every non-trivial normal subgroup of the corresponding split extension $C(p, q^{\infty})$ of A by Q contains A. It follows that the non-trivial normal subgroups of $C(p, q^{\infty})$ are in 1-1 inclusion preserving correspondence with the subgroups of Q, and because these subgroups are well ordered by inclusion, of order type $\omega + 1$, that $C(p, q^{\infty})$ satisfies Min-n and has no proper normal subgroups of finite index. This proves

(B) (Čarin, 1949). *There exist non-Abelian metabelian groups with* Min-n *and with no proper normal subgroups of finite index.*

The group described above is the analogue of (and is a union of) groups of importance in the theory of finite soluble groups. Groups which, like this group, have no proper (normal) subgroups of finite index are sometimes called *$\underline{\underline{F}}$-perfect groups.* Now any group G satisfying Min-n has a normal subgroup G_0 minimal with respect to having finite index in G. If H is a subgroup of finite index in G_0, then H must contain a normal subgroup of G of finite index, and, by definition of G_0, this subgroup must be G_0. So G_0 is $\underline{\underline{F}}$-perfect and is uniquely determined as the least subgroup of finite index in G. Therefore the second assertion in the next result follows from the first:

(C) (Wilson, 1970). *The property* Min-n *is inherited by normal subgroups of finite index. Every group satisfying* Min-n *is a finite extension of an $\underline{\underline{F}}$-perfect group satisfying* Min-n.

The significance of this result is that, for some purposes, it is sufficient to study $\underline{\underline{F}}$-perfect groups satisfying Min-n.

Work of Hartley and McDougall (1971) has shown that Čarin's example described earlier is rather typical among the $\underline{\underline{F}}$-perfect metabelian groups with Min-n; and in fact they have given a complete classification of all groups in a somewhat wider class of groups (a certain class of finite nilpotent extensions of $\underline{\underline{F}}$-perfect metabelian groups with Min-n). The starting point for the classification, so far as it concerns $\underline{\underline{F}}$-perfect groups, is the following basic result:

(D) (McDougall, 1970). *$\underline{\underline{F}}$-perfect metabelian groups with* Min-n *split* *conjugately over their derived groups; their maximal p-subgroups are Abelian, for each prime p.*

The classification theorem itself requires more notation for its precise statement than I want to introduce here; let it suffice to say that

(a) the derived group G' of an $\underline{F}$-perfect group G satisfying Min-n may be thought of as an Artinian module for $\mathbb{Z}X$, where X is a complement to G' in G, and as such is a direct sum of finitely many indecomposable submodules,

(b) every indecomposable Artinian $\mathbb{Z}X$-module is a p-group for some prime p, and is determined to isomorphism by its (possibly infinite) exponent and its unique minimal submodule,

(c) for each irreducible $\mathbb{Z}_pX$-module M, there are indecomposable $\mathbb{Z}X$-modules of exponent p^n for each integer $n > 0$, and of infinite exponent, having minimal submodules isomorphic to M, and

(d) irreducible $\mathbb{Z}_pX$-modules correspond to homomorphisms from X into the algebraic closure of $\mathbb{Z}_p$.

What I would like to do now is prove completely a result which, for metabelian groups, is a trivial consequence of the classification theorem of Hartley and McDougall.

(El) *(a) Artinian modules for countable Abelian groups are countable.*

(b) (McDougall, 1970). *Metabelian groups with* Min-n *are countable.*

(c) (Silcock, 1974). *Nilpotent by Abelian groups with* Min-n *are countable.*

The second and third statements here follow immediately from the first: if G is nilpotent by Abelian and satisfies Min-n, then each factor in the lower central series of the derived group G' of G can be considered as an Artinian module for the (countable) Abelian group G/G', so that G has a finite series with countable factors, and therefore is itself countable.

To prove the first statement, let A be a countable Abelian group and let $R = \mathbb{Z}A$. First let N be a cyclic Artinian R-module; then N is isomorphic as an R-module to R/I for some ideal I, and R/I is an Artinian ring (with identity), so has finite composition length as an R-module. Now let M be an arbitrary Artinian R-module and let M_i be the join of all submodules of composition length i, for each integer i. From above, we have $M = \cup M_i$. Moreover each $M_i + M_{i-1}/M_{i-1}$ is Artinian and a sum of irreducible modules, and therefore is countable because R is. It follows that each M_i is countable, and therefore that

M is countable.

The theory of Artinian rings and their representations provides a rather simple approach to the classification theorem of Hartley and McDougall for $\underline{F}$-perfect metabelian groups satisfying Min-n, if appeal is made to a result of Kovács and Newman (1962) and to the fundamental result (D). However, it seems that, to classify the groups in the wider class considered by Hartley and McDougall, one should proceed as they did, and use the theory of injective modules.

Before leaving metabelian groups, I am going to quote another result which can be read off from the classification theorem:

(E2) (Hartley and McDougall, 1971). *There are only countably many isomorphism classes of metabelian groups satisfying* Min-n.

Examples to show that there are $\underline{F}$-perfect soluble groups satisfying Min-n with arbitrarily large derived (and nilpotent) lengths were first constructed by McDougall (1970), and independently by Roseblade and Wilson (1971). Silcock (1974) gave examples of $\underline{F}$-perfect nilpotent by Abelian groups satisfying Min-n with arbitrarily large derived lengths. A useful testing-ground for conjectures of a general nature is provided by some other groups constructed by Silcock:

(F) (Silcock, 1975). *Let* $p_1, \ldots, p_n$ *be primes, with* $n \geq 2$ *and with* $p_i \neq p_{i+1}$ *and* $p_i - 1$ *indivisible by* p_{i+1} *for* $1 \leq i < n$; *and let* A_i *be a non-trivial cyclic or quasicyclic* p_i*-group for each* i. *There is a soluble group* G, *of derived length* n, *with the following properties:*

(a) the normal subgroups of G *are well ordered with respect to inclusion,*

(b) $G^{(i)}/G^{(i+1)}$ *is isomorphic to a direct power of* A_{n-i}, *for each* i, *and*

(c) G *is generated by isomorphic copies of the iterated (left normed) wreath products*

$$A_1 \text{ wr } A_3 \text{ wr } \ldots \text{ wr } A_{n-1+\varepsilon} \quad \textit{and} \quad A_2 \text{ wr } A_4 \text{ wr } \ldots \text{ wr } A_{n-\varepsilon}$$

where $\varepsilon = 0$ *if* n *is even and* $\varepsilon = 1$ *if* n *is odd.*

One would like to see many more constructions for soluble groups satisfying Min-n. I strongly suspect that none of the so far known examples give any measure of the complexity of arbitrary soluble groups satisfying Min-n.

Difficult problems arise when we try to prove results about general soluble or locally soluble groups satisfying Min-n. It would be unrealistic to hope for a classification of all soluble groups satisfying Min-n, or even all $\underline{F}$-perfect soluble groups satisfying Min-n - as unrealistic, for example, as to hope for a classification of all finitely generated soluble groups. The fact that $\underline{F}$-perfect metabelian groups with Min-n can be classified should be seen partly as a reflection of the strength of minimal conditions in ring theory. But there are few techniques for studying representations of infinite non-Abelian groups systematically, and it is with such representations that we are concerned when studying groups of derived length greater than two. The modules of interest to us are still Artinian, but the rings acting faithfully on them need not be right Artinian. One can proceed in either of two ways. One can look for fairly general results applicable to all soluble groups satisfying Min-n, or one can attempt to prove detailed results concerning small classes of groups. I would like to discuss these two approaches in turn.

The only completely general result known about soluble groups satisfying Min-n seems to be

(G) (Baer, 1964). *Soluble groups satisfying* Min-n *are locally finite.*

One feels that this result has some similarity to a result of Hall (1954), that soluble groups satisfying the maximal condition for normal subgroups are finitely generated. Murač (1973) has generalized Baer's result to show that *FC*-soluble groups satisfying Min-n are locally finite; and Hall's result may be similarly generalized. It is worth mentioning a still more general result for groups satisfying Min-n, if only because it leads to the fact that certain generalized soluble groups satisfying Min-n are actually soluble. Results of this kind seem to be rare.

(G*) (see Appendix). *(a) Let* G *be a group satisfying* Min-n *and having a finite series with locally Noetherian factors. If every chief factor of* G *is either Abelian or locally finite, then* G *is locally finite.*

(b) A group satisfying Min-n *is soluble if and only if it has a finite series with locally nilpotent factors.*

I have remarked that the metabelian groups satisfying Min-n are all countable, and that these groups fall into countably many isomorphism classes. Unfortunately there are no similar results for locally soluble groups, and Baer's theorem (G) does not extend to locally soluble groups. This follows from

(H) (Heineken and Wilson, 1974). *The class of locally soluble groups satisfying* Min-n *contains*

(a) groups with arbitrarily prescribed infinite cardinalities,

(b) $2^{\aleph_0}$ *isomorphism classes of countable groups, and*

(c) non-trivial torsion-free groups.

But this leaves countability questions for soluble groups, and they seem to me among the most interesting problems about general soluble groups with Min-n .

PROBLEM 1. *Are all soluble groups with* Min-n *countable?*

PROBLEM 2. *Do the soluble groups with* Min-n *fall into countably many isomorphism classes?*

Problem 1 was first raised in McDougall (1970)*. An affirmative answer would follow if it is the case that for every prime p and countable locally finite group G, the Z_pG-injective hull of each irreducible Z_pG-module is countable. On the other hand, it is easy to construct uncountable monolithic Artinian Z_pF-modules, where F is the free group of countably infinite rank. Nevertheless, the primitive estimates for cardinalities given by the usual embeddings of modules in injective modules do provide some information about the cardinalities of soluble groups satisfying Min-n (and therefore about the number of isomorphism classes of these groups).

(I) (see Appendix). *Let* G *be a soluble group with* Min-n *, of nilpotent length* n *; then the cardinality of* G *is at most* $\aleph_0$ *if* $n \leq 2$ *and otherwise at most*

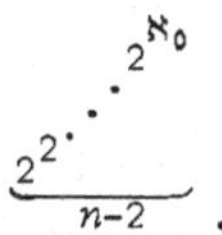

An attack on the countability problem for soluble groups of derived length three satisfying Min-n could perhaps be based on the classification theorem for $\underline{F}$-perfect metabelian groups. This shows that, if G is soluble of derived length three and if G satisfies Min-n , then G/G'' is a finite extension of a group much like Čarin's example (B). So as a first attempt to treat groups of length three, one might consider

PROBLEM 3. *Study irreducible and indecomposable modules over the prime field* Z_r *of* r *elements for Čarin's group* $C(p, q^\infty)$.

A solution (positive or negative) to Problem 1 would probably bring with it more detailed structural information, and might simultaneously answer the following question.

PROBLEM 4. *Is a soluble group satisfying* Min-n *necessarily locally an* r*-generator group, for some integer* r *? (The groups of nilpotent length at most two have this property.)*

Let me also mention a problem of a different nature.

PROBLEM 5. *Construct an infinite soluble group satisfying* Min-n *and having*

* Added January 1976. I have been informed that B. Hartley has recently constructed an example of an uncountable soluble group with Min-n .

no quasicyclic subgroups.

So much for general soluble groups satisfying Min-n . Now I want to talk about the alternative approach of considering a restricted class of soluble groups with Min-n , and asking for fairly detailed results. It does not seem clear to me where, if at all, one may expect to encounter problems here which are tractable and, at the same time, have the subtlety and depth of problems which have been studied concerning, say, finitely generated Abelian by nilpotent groups. These, by results of Hall (1954) are just the Abelian by nilpotent groups satisfying the *maximal* condition for normal subgroups. The tractability of those problems stemmed from the fact that they could in many cases be seen as problems in the representation theory of finitely generated nilpotent groups, and classical commutative algebra could be "twisted" to deal with such problems; and their depth stemmed from the fact that some of the "twists" required were very difficult. Nevertheless, commutative algebra provided inspiration and guidance.

Now Abelian by locally nilpotent groups satisfying Min-n are finite extensions of $\underline{\underline{F}}$-perfect metabelian groups satisfying Min-n (by the result of Čarin mentioned after (A)), so expectations that Abelian by locally nilpotent groups satisfying Min-n may behave in some respects like finitely generated Abelian by nilpotent groups can often be confirmed or denied by inspection of the classification theorem for $\underline{\underline{F}}$-perfect metabelian groups. For example, it is a fairly easy matter to check that *every metabelian by finite group satisfying* Min-n *has a bound on the lengths of chains of centralizers of its subgroups, while* $\underline{\underline{F}}$*-perfect centre by metabelian groups satisfying* Min-n *may have infinite chains of centralizers*; the proof by Lennox and Roseblade (1970) of corresponding results for finitely generated Abelian by nilpotent and centre by metabelian groups is very considerably more difficult. On the other hand, the representation theory of the metabelian groups satisfying Min-n appears so hard that one scarcely hopes for detailed results concerning the soluble groups of length three satisfying Min-n . One is led to consider classes of groups containing all the metabelian groups but not all the groups of length three. Candidates might be the class $\underline{\underline{X}}$ of nilpotent by Abelian by finite soluble groups satisfying Min-n , or the larger class of locally nilpotent by Abelian by finite soluble groups satisfying Min-n . (By (G*), this second class is a class of soluble groups.) Nevertheless, I have not been able to decide

PROBLEM 6. *Let* G *be an* $\underline{\underline{F}}$*-perfect locally nilpotent by Abelian group satisfying* Min-n . *(a) Is* G' *necessarily Abelian by nilpotent? (b) Is* G *necessarily countable?*

Some work has been done on the groups in the class $\underline{\underline{X}}$ just mentioned. The first result to appear concerned their Sylow structure:

(J) (Hartley and McDougall, 1971). *If* $H \le G \in \underline{\underline{X}}$ *, and if* π *is a set of primes, then the maximal* π*-subgroups of* H *are all conjugate.*

By contrast, Hartley (1971) remarked that $\underline{\underline{F}}$-perfect soluble groups of derived length three satisfying Min-n may have subgroups with non-conjugate maximal π-subgroups for suitable π .

It was shown by McDougall (1970) that $\underline{\underline{F}}$-perfect centre by metabelian groups satisfying Min-n need not split over their derived groups; however Silcock (1974) proved that *if* G *is an* $\underline{\underline{F}}$*-perfect nilpotent by Abelian group satisfying* Min-n , *then there is a divisible Abelian subgroup* X *such that* $G = G'X$ *and such that* $G' \cap X$ *lies in the centre of* G . So, at least if G has trivial centre, G splits over its derived group. This fact might make tractable the following problem, which is a special case of Problem 2:

PROBLEM 7. *Do the* $\underline{\underline{X}}$*-groups (or, more generally, the subgroups of* $\underline{\underline{X}}$*-groups) fall into countably many isomorphism classes?*

Because finite extensions present no difficulties, this is a question about nilpotent by Abelian groups.

Let me close with an eighth problem, concerning a concept which has been studied in relation to the class of finitely generated Abelian by nilpotent groups (Stroud, 1966) and various other classes. A group G is said to be *elliptic* for a word v if there is an integer n such that every element of the verbal subgroup $v(G)$ is a product of at most n v-values and inverses of v-values.

PROBLEM 8. *Are nilpotent by Abelian by finite groups satisfying* Min-n *elliptic for all words?*

Certainly metabelian by finite groups satisfying Min-n are elliptic for outer commutator words; in fact even nilpotent by metabelian by finite groups satisfying Min-n are elliptic for the word $[x, y]$.

APPENDIX

Here we indicate proofs of some of the results stated without proof in the lecture. We begin with the results (G*).

First we note that any group having a series

$$1 = G_0 \lhd \ldots \lhd G_n = G \qquad (*)$$

with each factor locally Noetherian also has such a series with $G_i \lhd G$ for each i ; this follows from the result of Baer (1957) that the class of locally Noetherian groups is closed with respect to normal products.

Now we take up the hypotheses of assertion (G*) (a). The above remark, and induction on the length of a series (*) with all factors locally Noetherian and all subgroups normal, allow us to restrict attention to a group G satisfying Min-n and having a locally Noetherian normal subgroup K with G/K locally finite.

Suppose that all chief factors of G below K are locally finite, and let T be a finitely generated subgroup of G. Then $T \cap K$ is finitely generated and $|T : T \cap K|$ is finite. If T were infinite we could find a subgroup $N \triangleleft G$, contained in K, minimal such that $T \cap N$ is finitely generated and infinite. Then N is finitely generated as a G-group (by the generators of $T \cap N$), and has a maximal proper G-subgroup N_1. Because N/N_1 is a chief factor of G, it must be locally finite, so that $T \cap N_1$ is finitely generated and infinite, a contradiction. Therefore T is finite, and G is locally finite.

It remains to show that each chief factor of G below K is locally finite. We suppose otherwise; then, passing to a homomorphic image, we can assume that G has a minimal normal subgroup M which is a torsion-free divisible Abelian group. Let $1 \neq m \in M$; then $M = \langle m^2 \rangle^G$, and $m \in \langle m^2 \rangle^F$ for a finitely generated subgroup F containing m. Because G/K is locally finite, $|F : F \cap K|$ is finite and $F \cap K$ is finitely generated; thus F is Noetherian. It follows that $\langle m \rangle^F$ is a non-trivial finitely generated torsion-free Abelian group but is equal to $\langle m^2 \rangle^F = \left(\langle m \rangle^F\right)^2$. With this contradiction, the proof of (G*) (a) is complete.

Now we suppose that G satisfies Min-n and has a finite series with locally nilpotent factors; we must prove that G is soluble. Arguing by induction on the length of an invariant series with locally nilpotent factors, we may suppose that G is a soluble extension of a locally nilpotent group. Because G satisfies Min-n, the derived series of G breaks off after finitely many terms in a perfect subgroup H, and H is locally nilpotent. Now G is locally finite by (G*) (a), and each locally nilpotent normal subgroup of a periodic locally soluble group centralizes all chief factors (*cf.* Gardiner, Hartley and Tomkinson, 1971). So H has an ascending series with all factors central; that is, H is hypercentral. But, by Grün's Lemma, non-trivial hypercentral groups cannot be perfect. Thus $H = 1$, and assertion (G*) (b) follows.

Next we turn to assertion (I). To prove this, it is enough to show that, if $H \triangleleft G$, and if H is nilpotent and satisfies the minimal condition on G-subgroups, then $|G| \leq 2^{\aleph}$, provided that $|G/H| = \aleph$ is infinite. Each factor group in the lower central series of H may be considered as a $\mathbb{Z}(G/H)$-module. If we show that every Artinian $\mathbb{Z}(G/H)$-module has cardinality at most $2^{\aleph}$, it will follow that G has a finite series with factors of cardinalities at most $2^{\aleph}$, and that $|G| \leq 2^{\aleph}$.

Let M be an Artinian $\mathbb{Z}\Gamma$-module, where $\Gamma = G/H$, and let S be the join of the minimal submodules of M. Then S is a direct sum of finitely many irreducible $\mathbb{Z}\Gamma$-modules, and so

$$|S| \le |Z\Gamma| = |\Gamma| .$$

However every non-trivial submodule of M has non-trivial intersection with S, so that M may be embedded in an injective hull of S, and therefore in $\mathrm{Hom}_Z(Z\Gamma, D)$ for any divisible Abelian group D containing S (*cf.* Curtis and Reiner (1962), p. 389). Since we may choose D with $|D| = \max\{|S|, \aleph_0\}$, it follows that

$$|M| \le |\mathrm{Hom}_Z(Z\Gamma, D)| = |D|^{|Z\Gamma|} = 2^{\aleph} ,$$

and (I) follows.

Finally, we sketch the proofs of our assertions concerning centralizer chains. The property of having all centralizer chains (finite and) of bounded length is inherited by subgroups, finite direct products, and finite extensions (see Houghton, Lennox and Wiegold, 1975). Thus, in discussing the boundedness of centralizer chains in classes of groups satisfying Min-n, it is enough to consider $\underline{\underline{F}}$-perfect monolithic groups.

Let G be an $\underline{\underline{F}}$-perfect monolithic metabelian group satisfying Min-n, and let X be a complement to G'. We may suppose $G' \ne 1$; then G' is the unique Sylow p-subgroup of G, for some prime p, and every p'-subgroup lies in some conjugate of X. Moreover the centre of G is trivial, and X acts faithfully on G'.

If $1 \ne T \le X$, then T acts fixed-point-freely on the unique minimal normal subgroup of G and so fixed-point-freely on G'. Thus $C_G(T) = X$. If $1 \ne T \le A$, then $C_G(T) = A$. If $T \nleq A$ and $T \nleq X^g$ for all $g \in G$, then T is neither a p-group nor a p'-group, so that $T \cap A \ne 1$ and $T \cap X^g \ne 1$ for some g; thus

$$C_G(T) \le A \cap X^g = 1 .$$

It follows immediately that any chain of centralizers in G has at most three terms.

We now exhibit an example of a centre by metabelian group satisfying Min-n and having infinite centralizer chains. Let p and q be distinct primes, and let F be the subfield of the algebraic closure of the field of p elements generated by the set Q of all qth power roots of unity. Consider the group G of 3×3 matrices over F of the form

$$\begin{pmatrix} x & 0 & 0 \\ r & 1 & 0 \\ s & t & x \end{pmatrix} \qquad (r, s, t \in F, x \in Q) .$$

The centre of G is the set of all matrices of G with $r = t = 0$ and $x = 1$. We let $\overline{G}$ be a homomorphic image of G with centre of order p; then $\overline{G}$ is an

$\underline{\underline{F}}$-perfect centre by metabelian group satisfying Min-n . Its derived group, the image in $\bar{G}$ of the group of lower triangular matrices, is an infinite extra-special p-group, and so has infinite centralizer chains.

BIBLIOGRAPHY

(This bibliography excludes the papers on locally soluble groups satisfying Min-n concerned primarily with conditions for such groups to be soluble Černikov groups. Many of these papers are discussed in Robinson (1972), Chapter 5.)

Reinhold Baer (1949), "Groups with descending chain condition for normal subgroups", *Duke Math. J.* **16**, 1-22. MR**10**,506.

Reinhold Baer (1957), "Lokal Noethersche Gruppen", *Math. Z.* **66** (1956/57), 341-363. MR**19**,12.

Reinhold Baer (1964), "Irreducible groups of automorphisms of abelian groups", *Pacific J. Math.* **14**, 385-406. MR**29**#2310.

В.С. Чарин [V.S. Čarin] (1949), "Замечание об условии минимальности для подгрупп" [A remark on the minimal condition for subgroups], *Dokl. Akad. Nauk SSSR (N.S.)* **66**, 575-576. MR**10**,677.

В.С. Чарин [V.S. Čarin] (1953), "Об условии минимальности для нормальных делителей локально разрешимых групп" [On the minimal condition for normal divisors of a locally soluble group], *Mat. Sb. (N.S.)* **33** (**75**), 27-36. MR**15**,197.

С.Н. Черников [S.N. Černikov] (1940), "Бесконечные локально разрешимые группы" [Infinite locally soluble groups], *Rec. Math.* [*Mat. Sb.*] *(N.S.)* **7** (**49**), 35-64. MR**2**,5.

С.Н. Черников [S.N. Černikov] (1949), "К теории локально разрешимых групп с условием минимальности для подгрупп" [On the theory of locally soluble groups with the minimal condition for subgroups], *Dokl. Akad. Nauk SSSR (N.S.)* **65**, 21-24. MR**10**,590.

Charles W. Curtis, Irving Reiner (1962), *Representation Theory of Finite Groups and Associative Algebras* (Pure and Applied Mathematics, **11**. Interscience [John Wiley & Sons], New York, London). MR**26**#2519.

A.D. Gardiner and B. Hartley and M.J. Tomkinson (1971), "Saturated formations and Sylow structure in locally finite groups", *J. Algebra* **17**, 177-211. MR**42**#7778.

P. Hall (1954), "Finiteness conditions for soluble groups", *Proc. London Math. Soc.* (3) **4**, 419-436. MR**17**,344.

B. Hartley (1971), "Sylow subgroups of locally finite groups", *Proc. London Math. Soc.* (3) **23**, 159-192. MR**46**#3623.

B. Hartley and D. McDougall (1971), "Injective modules and soluble groups satisfying the minimal condition for normal subgroups", *Bull. Austral. Math. Soc.* **4**, 113-135. MR43#2067.

H. Heineken and J.S. Wilson (1974), "Locally soluble groups with MIN-n ", *J. Austral. Math. Soc.* **17**, 305-318. MR50#7351.

C.H. Houghton, J.C. Lennox and James Wiegold (1975), "Centrality in wreath products", *J. Algebra* **35**, 356-366.

L.G. Kovács and M.F. Newman (1962), "Direct complementation in groups with operators", *Arch. der Math.* **13**, 427-433. MR26#5079.

J.C. Lennox and J.E. Roseblade (1970), "Centrality in finitely generated soluble groups", *J. Algebra* **16**, 399-435. MR42#4633.

David McDougall (1970), "Soluble groups with the minimum condition for normal subgroups", *Math. Z.* **118**, 157-167. MR46#7385.

D.H. McLain (1959), "Finiteness conditions in locally soluble groups", *J. London Math. Soc.* **34**, 101-107. MR21#2003.

Ю.М. Межебовский [Ju.M. Mežebovskiĭ] (1972), "О группах, обладающих возрастающим инвариантным рядом с конечными факторами" [On groups having an ascending invariant series with finite factors], *Sibirsk. Mat. Ž.* **13**, 473-476; *Siberian Math. J.* **13** (1972), 328-330. MR45#8738.

М.М. Мурач [M.M. Murač] (1973), "Про FC-розв'язні групи автоморфізмів розв'язних груп скінченного рангу" [FC-solvable groups of automorphisms of solvable groups of finite rank], *Dopovīdī Akad. Nauk Ukrain. RSR Ser. A* **1973**, 696-698; 764. MR48#4126.

Derek J.S. Robinson (1972), *Finiteness Conditions and Generalized Soluble Groups*, Part 1 (Ergebnisse der Mathematik und ihrer Grenzgebiete, **62**. Springer-Verlag, Berlin, Heidelberg, New York). MR48#11314.

J.E. Roseblade and J.S. Wilson (1971), "A remark about monolithic groups", *J. London Math. Soc.* (2) **3**, 361-362. MR43#7502.

Howard L. Silcock (1974), "Metanilpotent groups satisfying the minimal condition for normal subgroups", *Math. Z.* **135**, 165-173. MR51#711.

Howard L. Silcock (1975), "On the construction of soluble groups satisfying the minimal condition for normal subgroups", *Bull. Austral. Math. Soc.* **12**, 231-257. ZblM292#20027.

P.W. Stroud (1966), "Topics in the theory of verbal subgroups", Doctoral Dissertation, Cambridge, 1966.

John S. Wilson (1970), "Some properties of groups inherited by normal subgroups of finite index", *Math. Z.* **114**, 19-21. MR41#3577.

John S. Wilson (1974), "A note on subsoluble groups", *Proc. Second Internat. Conf. Theory of Groups*, Canberra, 1973, 717-718 (Lecture Notes in Mathematics, 372. Springer-Verlag, Berlin, Heidelberg, New York, 1974). Zbl M292#20029.

Department of Mathematics,
Australian National University,
Canberra, ACT, Australia;

Usual address:
Department of Pure Mathematics and Mathematical Statistics,
16 Mill Lane,
Cambridge CB2 1SB,
England.

20C15, 20C25

SOME RESULTS IN GROUP REPRESENTATION THEORY

G. Karpilovsky

Let G be a finite group and let K be an arbitrary field. Yamazaki ([4], Theorem 1) proved that there exists a finite central group extension of G by which all linearizable projective representations of G are linearized. This result motivates consideration of the following problem. Given a finite group G and an arbitrary field K of characteristic 0, what is the number of equivalence classes of irreducible linearizable projective representations of G over K? The following theorem gives the solution of this problem.

THEOREM 1. *Let* $(\hat{G}, \psi)$ *be the finite central group extension of* G *by which all the linearizable projective representations of* G *over the field* K *of characteristic* 0 *are linearized, and let* $A = \mathrm{Ker}\,\psi$. *Then the number of equivalence classes of irreducible linearizable projective representations of* G *over* K *is equal to the number of* K*-conjugacy classes of the group* $\hat{G}/A_K$ *which are in* $\hat{G}_K/A_K$. *Here* G_K *is the smallest subgroup containing* G' *such that* $\sqrt[n]{1} \in K$ *where* n *is the exponent of the group* G/G_K.

COROLLARY 1. *Let* K *be an algebraically closed field of characteristic* 0. *Then the number of equivalence classes of irreducible projective representations of* G *over* K *is equal to the number of conjugacy classes of the representation group* $\hat{G}$ *of* G *which are in* $\hat{G}'$.

COROLLARY 2. *Let* K *be the real number field. Then the number of equivalence classes of irreducible projective representations of* G *over* K *is equal to the number of* K*-conjugacy classes of the representation group* $\hat{G}$ *of* G *which are in* $\hat{G}_K$. *Here* $\hat{G}_K = \hat{G}$ *if* $2 \nmid (\hat{G} : \hat{G}')$ *and* $\hat{G}_K$ *is the minimal normal subgroup of* $\hat{G}$ *such that the factor-group* $\hat{G}/\hat{G}_K$ *is an elementary abelian* 2*-group if* $2 \mid (\hat{G} : \hat{G}')$.

Let G be a finite group and let K be an arbitrary field of characteristic

$p \geq 0$. The well-known result of Berman and Witt [1] states that the number of irreducible representations of G over K is equal to the number of K-conjugacy classes of p'-elements in G . The following theorem is a generalization of this.

THEOREM 2. *Let H be a normal subgroup of the finite group G , and let K be an arbitrary field of characteristic $p \geq 0$. Then the number of nonisomorphic KG-modules induced from the irreducible KH-modules is equal to the number of K-conjugacy classes of p'-elements of G which are in H .*

Let G be a finite group and let χ be an irreducible complex character of a normal subgroup N . If χ extends to a character of G then χ is stabilised by G , but the converse is false. The following theorem gives a sufficient condition for χ to be extended to a character of G .

THEOREM 3. *Let the group G contain a subgroup B of order n such that $G = N.B$ $(N \Delta G)$ and let χ be an irreducible complex character of N which is stabilised by G . Then χ extends to a character of G if the following conditions hold:*

(1) $(m, n) = 1$, $m = \chi(1)$;

(2) $N \cap B \leq N'$.

For the proof of Theorem 1 refer to [3]. Most of the notation we use is well known. In particular C^* denotes the set of all nonzero complex numbers, $O(g)$ is the order of g whilst I denotes the identity matrix.

Proof of Theorem 2. Let $T = \{\chi_1, \chi_2, \ldots, \chi_n\}$ and $S = \{K_1, K_2, \ldots, K_n\}$ be the sets of all irreducible K-characters of H and the set of all K-conjugacy classes of H respectively. The characters $\chi_1, \chi_2, \ldots, \chi_n$ are linearly independent over K and two irreducible KG-modules are isomorphic if and only if they have the same characters ([2], (30.12), (30.15), (29.7)). Using the same arguments as in Lemma 5 ([3]), it follows that if Γ_1 and Γ_2 are irreducible K-representations of H then the induced representations Γ_1^G and Γ_2^G of the group G are equivalent if and only if Γ_1 and Γ_2 are G-conjugate. Thus the matrix $M = \|\chi_i(K_j)\|$ $(1 \leq i, j \leq n)$ is invertible and the number of nonequivalent K-representations of G induced from irreducible K-representations of H is the same as the number r_1 of systems of transitivity in the permutation group

$$A = \left\{\begin{pmatrix}\chi_i \\ \chi_i^g\end{pmatrix} \;\middle|\; g \in G\right\} .$$

Using the same arguments as in Lemma 1 ([3]), it follows that the number of K-conjugacy classes of p'-elements of G which are in H is the same as the number

r_2 of systems of transitivity in the permutation group

$$B = \left\{ \begin{pmatrix} K_i \\ \\ K_i^g \end{pmatrix} \;\middle|\; K_i^g = g^{-1}K_i g ,\ g \in G \right\} .$$

Hence to prove the theorem it suffices to show that $r_1 = r_2$. Let A_g and B_g be the permutation matrices corresponding to the permutations $\pi_g = \begin{pmatrix} \chi_i \\ \\ \chi_i^g \end{pmatrix}$, $\tau_g = \begin{pmatrix} K_j \\ \\ K_j^g \end{pmatrix}$ respectively. For every $g \in G$ the permutation π_g applied to the rows of M and the permutation τ_g applied to the columns of M both map M to the same matrix $\left\|\chi_i^g(K_j)\right\|$ $(1 \le i, j \le n)$. Hence $MA_g = B_gM$ for every $g \in G$ and so $A_g = M^{-1}B_gM$. This shows that if $\{\pi_{g_1}, \pi_{g_2}, \ldots, \pi_{g_m}\}$ are all distinct elements of the group A, then $\{\tau_{g_1}, \tau_{g_2}, \ldots, \tau_{g_m}\}$ are all distinct elements of the group B. Thus $|A| = |B|$ and

$$\sum_{i=1}^{m} \operatorname{tr} A_{g_i} = \sum_{i=1}^{m} \operatorname{tr} B_{g_i} .$$

By ([2], (32.2)), $r_1 = r_2$, proving the theorem.

Proof of Theorem 3. Let Γ be a matrix representation of N which affords χ. Since χ is stabilised by G any two representations $s \to \Gamma(s)$ and $s \to \Gamma(g^{-1}sg)$ $(s \in N)$ of N are equivalent for all $g \in G$. Thus, if $g \in G$, there is a matrix $\psi(g)$ such that

$$\psi(g)^{-1}\Gamma(s)\psi(g) = \Gamma(g^{-1}sg) \quad \text{for any } s \in N \tag{1}$$

and so we may assume that

$$\Gamma(s) = \psi(s) \text{ , all } s \in N . \tag{2}$$

It is easy to see that the matrix $\psi(g_1)\psi(g_2)\psi(g_1g_2)^{-1}$ permutes with $\Gamma(s)$ for all $s \in N$, $g_1, g_2 \in G$, and thus it follows from the Schur's Lemma that

$$\psi(g_1)\psi(g_2) = \lambda(g_1, g_2)\psi(g_1g_2) \text{ , } \lambda(g_1, g_2) \in C^* . \tag{3}$$

Therefore ψ is a projective representation of G. By replacing $\psi(g)$ by $\delta_g\psi(g)$ for a suitable $\delta_g \in C^*$ we may assume that

$$\psi(g)^{O(g)} = I \quad \text{for any } g \in G\text{-}N . \tag{4}$$

If $g \in N \cap B$ then $\psi(g) = \Gamma(g)$ and so $N \cap B \leq N'$ implies that $\det \psi(g) = 1$. If $g \in B-N$ then it follows from (4) that $\det \psi(g) = \varepsilon$, $\varepsilon^k = 1$, $k = O(g)$. The condition $(m, n) = 1$ implies $(m, k) = 1$ and hence there exists a natural number x such that $mx \equiv 1 \pmod{k}$. Thus $\det \varepsilon^{-x}\psi(g) = 1$ and $[\varepsilon^{-x}\psi(g)]^{O(g)} = I$. We may therefore assume that

$$\det \psi(g) = 1 , \quad \psi(g)^{O(g)} = I \quad \text{for all } g \in B . \tag{5}$$

Calculating the determinants in (3) and applying (5) we obtain

$$[\lambda(g_1, g_2)]^m = 1 \quad \text{for all } g_1, g_2 \in B . \tag{6}$$

Now consider the group $L = \langle \lambda(g) \mid g \in B \rangle$. Then L contains the central subgroup $M = \langle \lambda(g_1, g_2)I \mid g_1, g_2 \in B \rangle$. It follows from (5), (6) and $(m, n) = 1$ that the factor-group L/M has the order prime to the order of M and hence by a theorem of Schur ([2], (7.5)) the group L is the direct product of its Hall subgroups. Since L is generated by the elements of orders prime to m it follows that $M = 1$ and

$$\psi(g_1)\psi(g_2) = \psi(g_1g_2) \quad \text{for all } g_1, g_2 \in B . \tag{7}$$

Let R be a transversal to $N \cap B$ in B. Thus each g in G has a unique representation $g = rt$, $r \in R$, $t \in N$. If $g_1 = r_1t_1$ is another element of G write $rr_1 = r_2t_2$ with $r_2 \in R$, $t_2 \in N \cap B$. Define $\tilde{\psi}(g) = \psi(r)\psi(t)$. Using (1), (2) and (7) we get $\tilde{\psi}(g)\tilde{\psi}(g_1) = \tilde{\psi}(gg_1)$. Thus the character afforded by $\tilde{\psi}$ extends χ. This completes the proof.

References

[1] С.Д. Берман [S.D. Berman], "Число неприводимых представлний конечной группы над произвольным прлем" [The number of irreducible representations of a finite group over an arbitrary field], *Dokl. Akad. Nauk SSSR (N.S.)* 106 (1956), 767-769. MR17,1181.

[2] Charles W. Curtis, Irving Reiner, *Representation Theory of Finite Groups and Associative Algebras* (Pure and Applied Mathematics, 11. Interscience [John Wiley & Sons], New York, London, 1962). MR26#2519.

[3] G. Karpilovsky, "On linearizable irreducible projective representations of finite groups", *J. Math. Soc. Japan* 28 (1976), 541-549.

[4] Keijiro Yamazaki, "A note on projective representations of finite groups", *Sci. Papers College Gen. Ed. Univ. Tokyo* 14 (1964), 27-36. MR30#3153.

School of Mathematics,
University of New South Wales, Kensington,
New South Wales, Australia.

Vol. 399: Functional Analysis and its Applications. Proceedings 1973. Edited by H. G. Garnir, K. R. Unni and J. H. Williamson. II, 584 pages. 1974.

Vol. 400: A Crash Course on Kleinian Groups. Proceedings 1974. Edited by L. Bers and I. Kra. VII, 130 pages. 1974.

Vol. 401: M. F. Atiyah, Elliptic Operators and Compact Groups. V, 93 pages. 1974.

Vol. 402: M. Waldschmidt, Nombres Transcendants. VIII, 277 pages. 1974.

Vol. 403: Combinatorial Mathematics. Proceedings 1972. Edited by D. A. Holton. VIII, 148 pages. 1974.

Vol. 404: Théorie du Potentiel et Analyse Harmonique. Edité par J. Faraut. V, 245 pages. 1974.

Vol. 405: K. J. Devlin and H. Johnsbråten, The Souslin Problem. VIII, 132 pages. 1974.

Vol. 406: Graphs and Combinatorics. Proceedings 1973. Edited by R. A. Bari and F. Harary. VIII, 355 pages. 1974.

Vol. 407: P. Berthelot, Cohomologie Cristalline des Schémas de Caracteristique p > o. II, 604 pages. 1974.

Vol. 408: J. Wermer, Potential Theory. VIII, 146 pages. 1974.

Vol. 409: Fonctions de Plusieurs Variables Complexes, Séminaire François Norguet 1970-1973. XIII, 612 pages. 1974.

Vol. 410: Séminaire Pierre Lelong (Analyse) Année 1972-1973. VI, 181 pages. 1974.

Vol. 411: Hypergraph Seminar. Ohio State University, 1972. Edited by C. Berge and D. Ray-Chaudhuri. IX, 287 pages. 1974.

Vol. 412: Classification of Algebraic Varieties and Compact Complex Manifolds. Proceedings 1974. Edited by H. Popp. V, 333 pages. 1974.

Vol. 413: M. Bruneau, Variation Totale d'une Fonction. XIV, 332 pages. 1974.

Vol. 414: T. Kambayashi, M. Miyanishi and M. Takeuchi, Unipotent Algebraic Groups. VI, 165 pages. 1974.

Vol. 415: Ordinary and Partial Differential Equations. Proceedings 1974. XVII, 447 pages. 1974.

Vol. 416: M. E. Taylor, Pseudo Differential Operators. IV, 155 pages. 1974.

Vol. 417: H. H. Keller, Differential Calculus in Locally Convex Spaces. XVI, 131 pages. 1974.

Vol. 418: Localization in Group Theory and Homotopy Theory and Related Topics. Battelle Seattle 1974 Seminar. Edited by P. J. Hilton. VI, 172 pages 1974.

Vol. 419: Topics in Analysis. Proceedings 1970. Edited by O. E. Lehto, I. S. Louhivaara, and R. H. Nevanlinna. XIII, 392 pages. 1974.

Vol. 420: Category Seminar. Proceedings 1972/73. Edited by G. M. Kelly. VI, 375 pages. 1974.

Vol. 421: V. Poénaru, Groupes Discrets. VI, 216 pages. 1974.

Vol. 422: J.-M. Lemaire, Algèbres Connexes et Homologie des Espaces de Lacets. XIV, 133 pages. 1974.

Vol. 423: S. S. Abhyankar and A. M. Sathaye, Geometric Theory of Algebraic Space Curves. XIV, 302 pages. 1974.

Vol. 424: L. Weiss and J. Wolfowitz, Maximum Probability Estimators and Related Topics. V, 106 pages. 1974.

Vol. 425: P. R. Chernoff and J. E. Marsden, Properties of Infinite Dimensional Hamiltonian Systems. IV, 160 pages. 1974.

Vol. 426: M. L. Silverstein, Symmetric Markov Processes. X, 287 pages. 1974.

Vol. 427: H. Omori, Infinite Dimensional Lie Transformation Groups. XII, 149 pages. 1974.

Vol. 428: Algebraic and Geometrical Methods in Topology, Proceedings 1973. Edited by L. F. McAuley. XI, 280 pages. 1974.

Vol. 429: L. Cohn, Analytic Theory of the Harish-Chandra C-Function. III, 154 pages. 1974.

Vol. 430: Constructive and Computational Methods for Differential and Integral Equations. Proceedings 1974. Edited by D. L. Colton and R. P. Gilbert. VII, 476 pages. 1974.

Vol. 431: Séminaire Bourbaki - vol. 1973/74. Exposés 436-452. IV, 347 pages. 1975.

Vol. 432: R. P. Pflug, Holomorphiegebiete, pseudokonvexe Gebiete und das Levi-Problem. VI, 210 Seiten. 1975.

Vol. 433: W. G. Faris, Self-Adjoint Operators. VII, 115 pages. 1975.

Vol. 434: P. Brenner, V. Thomée, and L. B. Wahlbin, Besov Spaces and Applications to Difference Methods for Initial Value Problems. II, 154 pages. 1975.

Vol. 435: C. F. Dunkl and D. E. Ramirez, Representations of Commutative Semitopological Semigroups. VI, 181 pages. 1975.

Vol. 436: L. Auslander and R. Tolimieri, Abelian Harmonic Analysis, Theta Functions and Function Algebras on a Nilmanifold. V, 99 pages. 1975.

Vol. 437: D. W. Masser, Elliptic Functions and Transcendence. XIV, 143 pages. 1975.

Vol. 438: Geometric Topology. Proceedings 1974. Edited by L. C. Glaser and T. B. Rushing. X, 459 pages. 1975.

Vol. 439: K. Ueno, Classification Theory of Algebraic Varieties and Compact Complex Spaces. XIX, 278 pages. 1975

Vol. 440: R. K. Getoor, Markov Processes: Ray Processes and Right Processes. V, 118 pages. 1975.

Vol. 441: N. Jacobson, PI-Algebras. An Introduction. V, 115 pages. 1975.

Vol. 442: C. H. Wilcox, Scattering Theory for the d'Alembert Equation in Exterior Domains. III, 184 pages. 1975.

Vol. 443: M. Lazard, Commutative Formal Groups. II, 236 pages. 1975.

Vol. 444: F. van Oystaeyen, Prime Spectra in Non-Commutative Algebra. V, 128 pages. 1975.

Vol. 445: Model Theory and Topoi. Edited by F. W. Lawvere, C. Maurer, and G. C. Wraith. III, 354 pages. 1975.

Vol. 446: Partial Differential Equations and Related Topics. Proceedings 1974. Edited by J. A. Goldstein. IV, 389 pages. 1975.

Vol. 447: S. Toledo, Tableau Systems for First Order Number Theory and Certain Higher Order Theories. III, 339 pages. 1975.

Vol. 448: Spectral Theory and Differential Equations. Proceedings 1974. Edited by W. N. Everitt. XII, 321 pages. 1975.

Vol. 449: Hyperfunctions and Theoretical Physics. Proceedings 1973. Edited by F. Pham. IV, 218 pages. 1975.

Vol. 450: Algebra and Logic. Proceedings 1974. Edited by J. N. Crossley. VIII, 307 pages. 1975.

Vol. 451: Probabilistic Methods in Differential Equations. Proceedings 1974. Edited by M. A. Pinsky. VII, 162 pages. 1975.

Vol. 452: Combinatorial Mathematics III. Proceedings 1974. Edited by Anne Penfold Street and W. D. Wallis. IX, 233 pages. 1975.

Vol. 453: Logic Colloquium. Symposium on Logic Held at Boston, 1972-73. Edited by R. Parikh. IV, 251 pages. 1975.

Vol. 454: J. Hirschfeld and W. H. Wheeler, Forcing, Arithmetic, Division Rings. VII, 266 pages. 1975.

Vol. 455: H. Kraft, Kommutative algebraische Gruppen und Ringe. III, 163 Seiten. 1975.

Vol. 456: R. M. Fossum, P. A. Griffith, and I. Reiten, Trivial Extensions of Abelian Categories. Homological Algebra of Trivial Extensions of Abelian Categories with Applications to Ring Theory. XI, 122 pages. 1975.

Vol. 457: Fractional Calculus and Its Applications. Proceedings 1974. Edited by B. Ross. VI, 381 pages. 1975.

Vol. 458: P. Walters, Ergodic Theory – Introductory Lectures. VI, 198 pages. 1975.

Vol. 459: Fourier Integral Operators and Partial Differential Equations. Proceedings 1974. Edited by J. Chazarain. VI, 372 pages. 1975.

Vol. 460: O. Loos, Jordan Pairs. XVI, 218 pages. 1975.

Vol. 461: Computational Mechanics. Proceedings 1974. Edited by J. T. Oden. VII, 328 pages. 1975.

Vol. 462: P. Gérardin, Construction de Séries Discrètes p-adiques. »Sur les séries discrètes non ramifiées des groupes réductifs déployés p-adiques«. III, 180 pages. 1975.

Vol. 463: H.-H. Kuo, Gaussian Measures in Banach Spaces. VI, 224 pages. 1975.

Vol. 464: C. Rockland, Hypoellipticity and Eigenvalue Asymptotics. III, 171 pages. 1975.

Vol. 465: Séminaire de Probabilités IX. Proceedings 1973/74. Edité par P. A. Meyer. IV, 589 pages. 1975.

Vol. 466: Non-Commutative Harmonic Analysis. Proceedings 1974. Edited by J. Carmona, J. Dixmier and M. Vergne. VI, 231 pages. 1975.

Vol. 467: M. R. Essén, The Cos $\pi\lambda$ Theorem. With a paper by Christer Borell. VII, 112 pages. 1975.

Vol. 468: Dynamical Systems – Warwick 1974. Proceedings 1973/74. Edited by A. Manning. X, 405 pages. 1975.

Vol. 469: E. Binz, Continuous Convergence on C(X). IX, 140 pages. 1975.

Vol. 470: R. Bowen, Equilibrium States and the Ergodic Theory of Anosov Diffeomorphisms. III, 108 pages. 1975.

Vol. 471: R. S. Hamilton, Harmonic Maps of Manifolds with Boundary. III, 168 pages. 1975.

Vol. 472: Probability-Winter School. Proceedings 1975. Edited by Z. Ciesielski, K. Urbanik, and W. A. Woyczyński. VI, 283 pages. 1975.

Vol. 473: D. Burghelea, R. Lashof, and M. Rothenberg, Groups of Automorphisms of Manifolds. (with an appendix by E. Pedersen) VII, 156 pages. 1975.

Vol. 474: Séminaire Pierre Lelong (Analyse) Année 1973/74. Edité par P. Lelong. VI, 182 pages. 1975.

Vol. 475: Répartition Modulo 1. Actes du Colloque de Marseille-Luminy, 4 au 7 Juin 1974. Edité par G. Rauzy. V, 258 pages. 1975. 1975.

Vol. 476: Modular Functions of One Variable IV. Proceedings 1972. Edited by B. J. Birch and W. Kuyk. V, 151 pages. 1975.

Vol. 477: Optimization and Optimal Control. Proceedings 1974. Edited by R. Bulirsch, W. Oettli, and J. Stoer. VII, 204 pages. 1975

Vol. 478: G. Schober, Univalent Functions – Selected Topics. V, 200 pages. 1975.

Vol. 479: S. D. Fisher and J. W. Jerome, Minimum Norm Extremals in Function Spaces. With Applications to Classical and Modern Analysis. VIII, 209 pages. 1975.

Vol. 480: X. M. Fernique, J. P. Conze et J. Gani, Ecole d'Eté de Probabilités de Saint-Flour IV–1974. Edité par P.-L. Hennequin. XI, 293 pages. 1975.

Vol. 481: M. de Guzmán, Differentiation of Integrals in R^n. XII, 226 pages. 1975.

Vol. 482: Fonctions de Plusieurs Variables Complexes II. Séminaire François Norguet 1974–1975. IX, 367 pages. 1975.

Vol. 483: R. D. M. Accola, Riemann Surfaces, Theta Functions, and Abelian Automorphisms Groups. III, 105 pages. 1975.

Vol. 484: Differential Topology and Geometry. Proceedings 1974. Edited by G. P. Joubert, R. P. Moussu, and R. H. Roussarie. IX, 287 pages. 1975.

Vol. 485: J. Diestel, Geometry of Banach Spaces – Selected Topics. XI, 282 pages. 1975.

Vol. 486: S. Stratila and D. Voiculescu, Representations of AF-Algebras and of the Group U (∞). IX, 169 pages. 1975.

Vol. 487: H. M. Reimann und T. Rychener, Funktionen beschränkter mittlerer Oszillation. VI, 141 Seiten. 1975.

Vol. 488: Representations of Algebras, Ottawa 1974. Proceedings 1974. Edited by V. Dlab and P. Gabriel. XII, 378 pages. 1975.

Vol. 489: J. Bair and R. Fourneau, Etude Géométrique des Espaces Vectoriels. Une Introduction. VII, 185 pages. 1975.

Vol. 490: The Geometry of Metric and Linear Spaces. Proceedings 1974. Edited by L. M. Kelly. X, 244 pages. 1975.

Vol. 491: K. A. Broughan, Invariants for Real-Generated Uniform Topological and Algebraic Categories. X, 197 pages. 1975.

Vol. 492: Infinitary Logic: In Memoriam Carol Karp. Edited by D. W. Kueker. VI, 206 pages. 1975.

Vol. 493: F. W. Kamber and P. Tondeur, Foliated Bundles and Characteristic Classes. XIII, 208 pages. 1975.

Vol. 494: A Cornea and G. Licea. Order and Potential Resolvent Families of Kernels. IV, 154 pages. 1975.

Vol. 495: A. Kerber, Representations of Permutation Groups II. V, 175 pages. 1975.

Vol. 496: L. H. Hodgkin and V. P. Snaith, Topics in K-Theory. Two Independent Contributions. III, 294 pages. 1975.

Vol. 497: Analyse Harmonique sur les Groupes de Lie. Proceedings 1973–75. Edité par P. Eymard et al. VI, 710 pages. 1975.

Vol. 498: Model Theory and Algebra. A Memorial Tribute to Abraham Robinson. Edited by D. H. Saracino and V. B. Weispfenning. X, 463 pages. 1975.

Vol. 499: Logic Conference, Kiel 1974. Proceedings. Edited by G. H. Müller, A. Oberschelp, and K. Potthoff. V, 651 pages 1975.

Vol. 500: Proof Theory Symposion, Kiel 1974. Proceedings. Edited by J. Diller and G. H. Müller. VIII, 383 pages. 1975.

Vol. 501: Spline Functions, Karlsruhe 1975. Proceedings. Edited by K. Böhmer, G. Meinardus, and W. Schempp. VI, 421 pages. 1976.

Vol. 502: János Galambos, Representations of Real Numbers by Infinite Series. VI, 146 pages. 1976.

Vol. 503: Applications of Methods of Functional Analysis to Problems in Mechanics. Proceedings 1975. Edited by P. Germain and B. Nayroles. XIX, 531 pages. 1976.

Vol. 504: S. Lang and H. F. Trotter, Frobenius Distributions in GL_2-Extensions. III, 274 pages. 1976.

Vol. 505: Advances in Complex Function Theory. Proceedings 1973/74. Edited by W. E. Kirwan and L. Zalcman. VIII, 203 pages. 1976.

Vol. 506: Numerical Analysis, Dundee 1975. Proceedings. Edited by G. A. Watson. X, 201 pages. 1976.

Vol. 507: M. C. Reed, Abstract Non-Linear Wave Equations. VI, 128 pages. 1976.

Vol. 508: E. Seneta, Regularly Varying Functions. V, 112 pages. 1976.

Vol. 509: D. E. Blair, Contact Manifolds in Riemannian Geometry. VI, 146 pages. 1976.

Vol. 510: V. Poènaru, Singularités C^∞ en Présence de Symétrie. V, 174 pages. 1976.

Vol. 511: Séminaire de Probabilités X. Proceedings 1974/75. Edité par P. A. Meyer. VI, 593 pages. 1976.

Vol. 512: Spaces of Analytic Functions, Kristiansand, Norway 1975. Proceedings. Edited by O. B. Bekken, B. K. Øksendal, and A. Stray. VIII, 204 pages. 1976.

Vol. 513: R. B. Warfield, Jr. Nilpotent Groups. VIII, 115 pages. 1976.

Vol. 514: Séminaire Bourbaki vol. 1974/75. Exposés 453 – 470. IV, 276 pages. 1976.

Vol. 515: Bäcklund Transformations. Nashville, Tennessee 1974. Proceedings. Edited by R. M. Miura. VIII, 295 pages. 1976.